T0155756

SpringerBriefs in Research Synthesis and Meta-Analysis

Series editors

Michael Bosnjak, Mannheim, Germany
Mike W.L. Cheung, Singapore, Singapore
Wolfgang Viechtbauer, Maastricht, The Netherlands

More information about this series at http://www.springer.com/series/10240

Suzanne Jak

Meta-Analytic Structural Equation Modelling

 Springer

Suzanne Jak
Faculty of Social and Behavioural Sciences
Utrecht University
Utrecht
The Netherlands

Part of this work was supported by Rubicon-grant 446-14-003 from the Netherlands Organization for Scientific Research (NWO).

ISSN 2193-0015 ISSN 2193-0023 (electronic)
SpringerBriefs in Research Synthesis and Meta-Analysis
ISBN 978-3-319-27172-9 ISBN 978-3-319-27174-3 (eBook)
DOI 10.1007/978-3-319-27174-3

Library of Congress Control Number: 2015957127

Printed on acid-free paper

This Springer imprint is published by SpringerNature
The registered company is Springer International Publishing AG Switzerland

Preface

Meta-analysis (MA) is a prominent statistical tool in many research disciplines. It is a statistical method to combine the data of several independent studies, in order to draw overall conclusions based on the pooled data. Structural equation modeling (SEM) is a technique that tests the relations between a set of variables in one model, for example in a path model or a factor model. In a SEM analysis, all hypothesized relations between the variables are tested simultaneously. The overall fit of the model can be evaluated using several fit indices. SEM does not need raw data, but fits structural equation models to covariance (or correlation) matrices directly.

The combination of meta-analysis and structural equation modeling for the purpose of testing hypothesized models is called meta-analytic structural equation modeling (MASEM). MASEM is a new and promising field of research. With MASEM, a single model can be tested to explain the relationships between a set of variables in several studies. By using MASEM, we can profit from all available information from all available studies, even if few or none of the studies report about all relationships that feature in the full model of interest.

I use the term MASEM for the process of fitting a structural equation model on the combined data from several studies. SEM can also be used to perform ordinary meta-analysis (SEM-based meta-analysis), but this falls outside the scope of this book.

This book gives an overview of the most prominent methods to perform MASEM, with a focus on the two-stage SEM approach. The fixed and the random approach to MASEM are illustrated with two applications to real data. All steps that have to be taken to perform the analyses are discussed. The data and syntax files can be found online (http://suzannejak.nl/masem), so that readers can replicate all analyses.

I would like to thank the editors of the Springer Briefs Series on Research Synthesis and Meta-Analysis, Mike Cheung, Michael Bosnjak, and Wolfgang Viechtbauer, for inviting me to write this book and providing me with valuable comments on earlier versions of the manuscript. Of course, all remaining errors are mine. I also thank Mathilde Verdam for providing feedback on the first chapter, and Debora Roorda and Huiyong Fan for making their data available.

I am especially grateful to Mike Cheung, who was willing to share his extensive knowledge of MASEM with me during my stay at the National University of Singapore.

September 2015   Suzanne Jak

Contents

Chapter 1
Introduction to Meta-Analysis and Structural Equation Modeling

Abstract Meta-analysis is a prominent statistical tool in many research disciplines. It is a statistical method to combine the effect sizes of separate independent studies, in order to draw overall conclusions based on the pooled results. Structural equation modeling is a multivariate technique to fit path models, factor models, and combinations of these to data. By combining meta-analysis and structural equation modeling, information from multiple studies can be used to test a single model that explains the relationships between a set of variables or to compare several models that are supported by different studies or theories. This chapter provides a short introduction to meta-analysis and structural equation modeling.

Keywords Meta-analysis · Introduction · Structural equation modeling · Path model · Factor model · Model fit

1.1 What Is Meta-Analysis?

The term "meta-analysis" was introduced by Glass (1976), who differentiated between primary analysis, secondary analysis, and meta-analysis. However, the techniques on which meta-analysis is based were developed much earlier (see Chalmers et al. 2002; O'Rourke 2007). In the terminology of Glass, primary analysis involves analyzing the data of a study for the first time. Secondary analysis involves the analysis of data that have been analyzed before, for example to check the results of previous analyses or to test new hypotheses. Meta-analysis then involves integration of the findings from several independent studies, by statistically combining the results of the separate studies. One of the first meta-analyses in the social sciences was performed by Smith and Glass (1977), who integrated the findings of 375 studies that investigated whether psychotherapy was beneficial for patients, a topic that was much debated at the time. By using a quantitative approach to standardizing and averaging treatment/control differences across all the studies, it appeared that overall, psychotherapy was effective, and that there is little difference in effectiveness across the different types of therapy. Around the same time as Smith and Glass performed this meta-analysis, other researchers developed similar techniques to synthesize research findings (Rosenthal and Rubin 1978, 1982; Schmidt and Hunter 1977), which are now all referred to as meta-analysis techniques. Meta-analysis is used to integrate findings in many fields, such as psychology, economy, education, medicine, and criminology.

1.1.1 Issues in Meta-Analysis

Compared with primary analysis, meta-analysis has important advantages. Because more data is used in a meta-analysis, the precision and accuracy of estimates can be improved. Increased precision and accuracy also leads to greater statistical power to detect effects.

Despite the obvious positive contributions of meta-analysis, the technique is also criticized. Sharpe (1997) identified the three main validity threats to meta-analysis: mixing of dissimilar studies, publication bias, and inclusion of poor quality studies. The mixing of dissimilar studies, also referred to as "mixing apples and oranges" problem, entails the issue that average effect sizes are not meaningful if they are aggregated over a very diverse range of studies. Card (2012) counters this critique by stating that it depends on the inference goal whether it is appropriate to include a broad range of studies in the meta-analysis (e.g. if one is interested in fruit, it is appropriate to include studies about apples, oranges, strawberries, banana's etc.). Moreover, a meta-analysis does not only entail aggregation across the total pool of studies, but can also be used to compare different subsets of studies using moderator analysis. The second threat, publication bias, is also referred to as the "file drawer" problem, and points to the problem that some studies that have been conducted may not be published, and are therefore not included in the meta-analysis. Publication bias is a real source of bias, because the non-published studies are probably those that found non-significant or unexpected results. Several methods exist that aim at avoiding, detecting and/or correcting for publication bias (see Rothstein et al. 2005; van Assen et al. 2014) but there is no consensus on the best ways to deal with the problem. The third issue, the inclusion of poor quality studies in the meta-analysis is also denoted as the "garbage in, garbage out" problem. Although it may seem logical to leave studies of poor quality out of the meta-analysis a priori, it is recommended to code the relevant features of the included primary studies that are required for high quality (e.g. randomization in an experiment), so that later on one can investigate whether these quality-conditions are related to the relevant effect sizes (Valentine 2009).

Cooper and Hedges (2009) distinguish six phases of research synthesis: Problem formulation, literature search, data evaluation, data analysis, interpretation of the results and presentation of the results. In this book we focus on the data analysis phase, referred to as meta-analysis. The other parts of research synthesis are discussed in for example Borenstein et al. (2009), Card (2012), Cooper et al. (2009), and Lipsey and Wilson (2001).

1.1.2   Statistical Analysis

Usually, the units of analysis in a meta-analysis are not the raw data, but summary statistics (effect size statistics) that are reported in the individual studies. The type of effect size statistic that is investigated depends on the nature of the variables involved. For example, if the interest is in differences between a treatment and control group on some continuous outcome variable, the meta-analysis may focus on the standardized mean difference (like Cohen's d or Hedges' g). If the hypothesis is about the association between two continuous variables, the (z-transformed) product moment correlation coefficient may be the focus of the analysis. If the interest is in association between two dichotomous variables, the (logged) odds ratio is often an appropriate effect size statistic. Once the effect size statistics of interest are gathered or reconstructed from the included studies, the statistical analysis can start, using fixed effects or random effects analysis.

The fixed effects approach is useful for conditional inference, which means that the conclusions cannot be generalized beyond the studies included in the analysis (Hedges and Vevea 1998). In the most common fixed effects model, it is assumed that the effect size statistics gathered from the studies are estimates of one population effect size, and differences between studies are solely the result of sampling error. The analysis focuses on obtaining a weighted mean effect size across studies. The weights are based on the sampling variance in the studies, so that studies with larger sampling variance (and smaller sample size) contribute less to the weighted mean effect size (which is the estimate of the population effect size).

The random effects approach facilitates inferences to studies beyond the ones included in the particular meta-analysis (unconditional inference). The random effects approach assumes that the population effect sizes vary from study to study, and that the studies in the meta-analysis are a random sample of studies that could have been included in the analysis. Differences in effect sizes between studies are hypothesized to be due to sampling error and other causes, such as differences in characteristics of the respondents or operationalization of the variables in the different studies. The random effects analysis leads to an estimate of the mean and variance of the distribution of effect sizes in the population.

Apart from the average effect size, it is often also of interest if and why studies differ systematically in their effect size statistics. Therefore, researchers often code study characteristics (e.g. average age of respondents, measurement instruments used, country in which the study was conducted), and investigate whether the effect sizes are associated with these study-level variables. This is called moderator analysis, and is used to investigate whether the association between the variables of interest is moderated by study characteristics. These moderator variables may explain variability in the effect sizes. If all variability is explained, a fixed effects model may hold, implying that conditional on the moderator variables, all remaining variability is sampling variability. If effect sizes are regressed on study level variables in a random effects approach, reflecting that the moderator variables do not explain all variability across the studies, this is called mixed effects meta-analysis.

To be consistent with recent terminology, I use the term "fixed effects model" for all models that do not estimate between-studies variance. This terminology is common in meta-analysis, but not in line with the statistical literature, where the fixed effects model denotes the model in which heterogeneity is explained by study-level variables. The model that assumes homogeneity of effect sizes, without study-level variables, is also called the "equal effects model" (Laird and Mosteller 1990). I use the term "fixed effects model" for both these models, and will explicitly state when study-level variables are included in the model.

1.2 What Is SEM?

Structural equation modeling (SEM) has roots in two very different techniques developed in two very different fields. Path analysis with its graphical representations of effects and effect decomposition comes from genetics research, where Wright (1920) proposed a method to predict heritability of the piebald pattern of guinea-pigs. Factor analysis is even older, with an early paper by Spearman (1904), and was developed in research on intelligence, to explain correlations between various ability tests (Spearman 1928). Jöreskog (1973) coined the name LISREL (LInear Structural RELations) for the framework that integrates the techniques of path analysis and factor analysis, as well as for the computer program that made the technique available to researchers.

1.2.1 Path Analysis

SEM is a confirmatory technique, which means that a model is formulated based on theory, and it is judged whether this model should be rejected by fitting the model to data. If multivariate normality of the data holds, the variance covariance matrix of the variables of interest and the sample size are sufficient to fit models to the data. This is a very convenient aspect of SEM, because it means that as long as authors report correlations and standard deviations of their research variables in their articles, other researchers are able to replicate the analyses, and to test different hypotheses on these data. In order to test hypotheses, these hypotheses have to be translated in a statistical model. The statistical model can be formulated in different ways, for example using a graphical display. The graphical displays that are used for structural equation models use squares to represent observed variables, ellipses to represent latent variables, one-headed arrows to represent regression

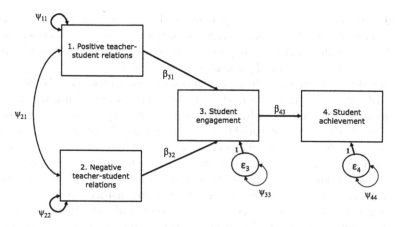

Fig. 1.1 Hypothesized path model in which the effects of Positive and Negative relations on achievement is fully mediated by engagement

coefficients, and two-headed arrows to represent covariances. Consider the path model in Fig. 1.1, in which the effect of negative and positive relations with teachers is hypothesized to affect student achievement through student engagement.

The four observed variables are depicted in squares. Student engagement is regressed on Positive and Negative relations, and Student Achievement is regressed on Student Engagement. There are no direct effects of Positive and Negative relations on Student Achievement, reflecting the hypothesis that these effects are fully mediated by Student Engagement. In this model, Engagement and Achievement are called endogenous variables, reflecting that other variables are hypothesized to have an effect on them. Variables that are not regressed on other variables are called exogenous variables. Positive and Negative relations are exogenous variables in this model. The two exogenous variables are assumed to covary, indicated by the two-headed arrow between them. There are also two-headed arrows pointing from the variable to itself, reflecting the variance of the variable (a covariance with itself is equal to a variance). The endogenous variables have a latent variable with variance pointing to it. This latent variable is called a residual factor, and could be viewed as a container variable representing all other variables that also explain variance in the endogenous variable, but that are not included in the model. The regression coefficient of the variable on the residual factor is not estimated but fixed at 1 for identification of the model. The variance of the residual factor represents the unexplained variance of the endogenous variable. So, part of the variance in Student Engagement is explained by Positive and Negative relations, and the remaining variance is residual variance (or, unexplained variance). Similarly, part of the variance in Student Achievement is explained by Student Engagement, and the remaining variance is residual variance. For the exogenous variables, actually, all variance is unexplained. So it seems logical to depict two more residual factors with variance pointing to Negative and Positive relations, instead of the double headed arrow pointing to the variables themselves. Indeed, this would be correct, but to keep the graphs simple they are often not depicted. Actually, the residual factor pointing to an endogenous variables is also often not fully depicted, but represented by a small one-sided arrow.

Attached to the arrows in the graphical display, the Greek symbols represent the model parameters. In a path model, the direct effects are often denoted by β and variances and covariances by ψ. For example, β_{43} represents the regression coefficient of Variable 4 on Variable 3, ψ_{44} represents the residual variance of Variable 4, and ψ_{21} represents the covariance between Variable 1 and Variable 2. The model parameters are collected in matrices. A path model on observed variables can be formulated using two matrices with parameters, matrix $\mathbf{B}$ and matrix Ψ, and an identity matrix, $\mathbf{I}$. For the example, these matrices look as follows, with rows 1–4 and columns 1–4 corresponding to the variables Positive relations, Negative relations, Student Achievement, and Student Engagement, respectively:

$$
\mathbf{B} = \begin{bmatrix} 0 & 0 & 0 & 0 \\ 0 & 0 & 0 & 0 \\ \beta_{31} & \beta_{32} & 0 & 0 \\ 0 & 0 & \beta_{43} & 0 \end{bmatrix}, \quad
\Psi = \begin{bmatrix} \psi_{11} & & & \\ \psi_{21} & \psi_{22} & & \\ 0 & 0 & \psi_{33} & \\ 0 & 0 & 0 & \psi_{44} \end{bmatrix} \text{and} \quad
\mathbf{I} = \begin{bmatrix} 1 & 0 & 0 & 0 \\ 0 & 1 & 0 & 0 \\ 0 & 0 & 1 & 0 \\ 0 & 0 & 0 & 1 \end{bmatrix}.
$$

Matrix Ψ is a symmetrical matrix, so the covariance between Variables 1 and 2 is equal to the covariance between Variable 2 and 1. Using these parameters, a model implied covariance matrix (Σ_{model}) can be formulated. The model implied covariance matrix is a function of the matrices with parameters:

$$
\Sigma_{\text{model}} = (\mathbf{I} - \mathbf{B})^{-1} \Psi (\mathbf{I} - \mathbf{B})^{-1T}. \tag{1.1}
$$

The resulting model implied covariance matrix (Σ_{model}) for the current example can be found in Appendix A. The basic hypothesis that is tested by fitting a structural equation model to data is:

$$
\Sigma = \Sigma_{\text{model}}. \tag{1.2}
$$

Note however, that the population covariance matrix, Σ, is generally unavailable to the researcher, who only observed a covariance matrix based on a sample, denoted $\mathbf{S}$. Suppose that observed covariance matrix of the four variables based on 104 respondents is as given in Table 1.1.

Table 1.1 Variances (on diagonal) and covariances of four research variables, $N = 104$

Variable	1	2	3	4
1. Positive relations	0.81			
2. Negative relations	−0.36	1.21		
3. Engagement	0.63	−0.60	1.69	
4. Achievement	0.14	−0.33	0.50	1.44

The model parameters that make up Σ_{model} can be estimated by minimizing a discrepancy function. This means that parameters are estimated in order to minimize the difference between the model implied covariance matrix (Σ_{model}), and the observed covariance matrix (**S**). The more parameters a model has, the easier it is to make the Σ_{model} close to **S**. The maximum number of parameters that a model can have in order to be identified is equal to the number of observed variances and covariances in **S**. In our example with four variables, the number of variances and covariances is ten. The number of parameters in the Σ_{model} equals eight (three regression coefficients, one covariance, four variances). The degrees of freedom (df) of a model are equal to the difference between these two. This model has 2 degrees of freedom. The larger the degrees of freedom of a model is, the more the model is a simplification of reality. Simple models are generally preferred over complicated models. But, the larger the degrees of freedom, the larger the difference between Σ_{model} and **S** will be, meaning that the absolute fit of a model will be worse.

Having less parameters than observed variances and covariances is not the only requirement for identification of the model. For a model to be identified, all parameters in the model need to be identified. See Bollen (1989) for an overview of methods to assess the identification of model parameters. If a model is identified, the parameters can be estimated. The most used estimation method is maximum likelihood (ML) estimation. The discrepancy function F_{ML} that is minimized with ML estimation is:

$$F_{\text{ML}} = \log |\Sigma_{\text{model}}| - \log |S| + \text{trace}\left(S\Sigma_{\text{model}}^{-1}\right) - p, \qquad (1.3)$$

where p is the number of variables in the model. If the model fits the data perfectly, the model implied covariance matrix will be equal to **S**, and F_{ML} will be zero. If the model does not fit perfectly, F_{ML} will be larger than zero. See Bollen (1989) for a description of ML and other estimation methods and their assumptions.

1.2.2 Model Fit

An important property of the ML estimator is that it provides a test of overall model fit for models with positive degrees of freedom. Under the null hypothesis ($\Sigma = \Sigma_{\text{model}}$), the minimum F_{ML} multiplied by the sample size minus one ($n - 1$) asymptotically follows a chi-square distribution, with degrees of freedom equal to the number of non-redundant elements in **S** minus the number of model parameters. If the chi-square value of a model is considered significant, the null hypothesis is rejected. The chi-square of a model may become significant because the discrepancy between **S** and the estimated Σ_{model} is large, or because the sample is large. With a very large sample, small differences between **S** and the estimated Σ_{model} may lead to a significant chi-square, and thus rejection of the model. Other fit measures are available in SEM, which do not test exact fit of the model, but are based on the idea that models are simplifications of reality and will never exactly hold in the population. The Root Mean Squared Error of Approximation (RMSEA, Steiger and Lind 1980) is the most prominent fit measure next to the chi-square. The RMSEA is interpreted using suggested cut-off values that should be regarded as rules of thumb. RMSEA values smaller than 0.05 are considered to indicate close fit, values smaller than 0.08 are considered satisfactory and values over 0.10 are considered indicative of bad fit (Browne and Cudeck 1992). Another prominent fit measure is the Comparative Fit Index (CFI, Bentler 1990) that is based on a comparison of the hypothesized model with the "independence model", which is a model in which all variables are unrelated. CFI values over 0.95 indicate reasonably good fit. For an overview of these and other fit indices see Schermelleh-Engel et al. (2003).

Fitting the model from Fig. 1.1 to the observed covariance matrix in Table 1.1 gives the following fit indices: $\chi^2 = 2.54$, df $= 2$, $p = 0.28$, RMSEA $= 0.05$ and CFI $= 0.99$. So, exact fit of the model is not rejected, and the model also fitted the data according to the rules of thumb for the RMSEA and CFI. If the model fits the data, the parameter estimates can be interpreted. If a model does not fit the data, the parameter estimates should not be interpreted because they will be wrong. Table 1.2 gives an overview of the unstandardized parameter estimates, the 95 % confidence intervals and the standardized parameter estimates of the model. See Appendix B for an example of an OpenMx-script to fit the current model.

All parameters in this model differ significantly from zero, as judged by the 95 % confidence intervals. For interpretation, it is useful to look at the standardized parameter estimates. For example, the standardized β_{31}, means that 1 standard deviation increase in Positive relationships is associated with 0.45 standard deviations increase in Engagement, controlled for the effect of Negative relationships. The standardized residual variance is interpreted as the proportion of residual

Table 1.2 Unstandardized parameter estimates, 95 % confidence intervals and standardized parameter estimates of the path model from Fig. 1.1

Parameter	Unstandardized estimate	95 % confidence interval		Standardized estimate
		Lower bound	Upper bound	
β_{31}	0.64	0.40	0.89	0.45
β_{32}	-0.30	-0.50	-0.10	-0.26
β_{43}	0.30	0.13	0.47	0.32
ψ_{21}	-0.36	-0.60	-0.36	-0.36
ψ_{11}	0.81	0.62	1.08	1.00
ψ_{22}	1.21	0.93	1.61	1.00
ψ_{33}	1.10	0.85	1.47	0.65
ψ_{44}	1.29	0.99	1.72	0.90
$\beta_{31} \times \beta_{43}$	0.19	0.08	0.34	0.14
$\beta_{32} \times \beta_{43}$	-0.09	-0.19	-0.03	-0.08

variance. This means that in the standardized solution, the proportion of explained variance in Student achievement is calculated as $1 - \psi_{44}, = 0.10$. The proportion of explained variance in Engagement is 0.35. Indirect effects are calculated as the product of the two direct effects that constitute the indirect effect. With OpenMx, one can estimate confidence intervals for indirect effects as well. The indirect effects of Positive and Negative relationships on Student Achievement are both small but significant (see the last two rows in Table 1.2). This shows that as expected, there is significant mediation. Whether there is full or partial mediation can be investigated by testing the significance of the direct effects of Positive and Negative relationships on Student Achievement. This is shown in Chap. 5.

1.2.3 Factor Analysis

Factor analysis can also be seen as a special case of structural equation modeling. Factor models involve latent variables that explain the covariances between the observed variables. Consider the two-factor model on five scales measuring children's problem behavior depicted in Fig. 1.2.

In a factor model, each indicator is affected by a common factor that explains the covariances between the indicators. The regression coefficients linking the factor to an indicator are called factor loadings. The larger a factor loading is, the more variance the factor explains in the indicator. Not all indicator variance may be common variance, which is reflected by the residual factors that affect each indicator. The variance of these residual factors is called residual variance (denoted by θ) and is assumed to consist of random error variance and structural

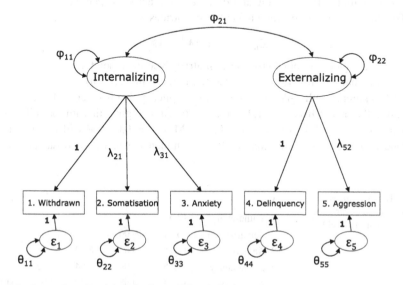

Fig. 1.2 A two-factor model on the five problem behavior variables

variance. For example, there may be a structural component in Somatisation that is not correlated with Anxiety or Withdrawn behavior.

With factor analysis, Σ_{model} is a function of factor loadings, depicted by λ's, factor variances and covariances, depicted by φ's, and residual variances, depicted by θ's. Note that one factor loading for each factor is fixed at 1. This is needed to identify the model. As factors are unobserved variables, the scale of the variables is not known, and a metric has to be given to the factors by fixing one factor loading per factor. Alternatively, one can fix the factor variances φ_{11} and φ_{22} at some value (e.g. 1) and estimate all factor loadings. In advanced models (e.g. multigroup and longitudinal models) one method of scaling may be preferred over the other, but in this example it is arbitrary how the factors are given a metric. The unstandardized parameters will differ based on the scaling method, but the model fit and the standardized parameter estimates will not. The factor model can be represented by three matrices with parameters, a full matrix Λ with factor loadings, a symmetrical matrix Φ with factor variances and covariances, and a symmetrical matrix Θ with residual variances and covariances. For the current model, the three matrices look as follows.

$$
\Lambda = \begin{bmatrix} 1 & 0 \\ \lambda_{21} & 0 \\ \lambda_{31} & 0 \\ 0 & 1 \\ 0 & \lambda_{52} \end{bmatrix}, \quad \Phi = \begin{bmatrix} \varphi_{11} & \\ \varphi_{21} & \varphi_{22} \end{bmatrix} \text{ and } \Theta = \begin{vmatrix} \theta_{11} & & & & \\ 0 & \theta_{22} & & & \\ 0 & 0 & \theta_{33} & & \\ 0 & 0 & 0 & \theta_{44} & \\ 0 & 0 & 0 & 0 & \theta_{55} \end{vmatrix}.
$$

The rows of Λ are associated with variables 1 through 5 from Fig. 1.2, as well as the rows and columns of Θ. The columns of Λ and the rows and columns of Φ are associated with the Internalizing and Externalizing factors respectively.

The factor model is specified using these matrices as:

$$
\Sigma_{model} = \Lambda \Phi \Lambda^{T} + \Theta, \tag{1.4}
$$

leading to the model implied covariance matrix given in Appendix C.

Suppose that we observed the covariance matrix of the five variables from a sample of 155 parents with children suffering from epilepsy that is given in Table 1.3.

Fitting the model from Fig. 1.2 to these data leads to good fit with the following fit measures: $\chi^2 = 4.08$, df $= 4$, $p = 0.40$, RMSEA $= 0.01$ and CFI $= 1.00$. The unstandardized parameter estimates, 95 % confidence intervals and standardized

Table 1.3 Variances (on diagonal) and covariances of five research variables

Variable	1	2	3	4	5
1. Withdrawn	12.55				
2. Somatization	6.31	10.06			
3. Anxiety	11.15	9.64	26.02		
4. Delinquency	2.85	2.09	4.84	3.72	
5. Aggression	12.44	9.68	22.20	9.96	51.02

Table 1.4 Unstandardized parameter estimates, 95 % confidence intervals and standardized parameter estimates of the factor model from Fig. 1.2

Parameter	Unstandardized estimate	95 % confidence interval		Standardized estimate
		Lower bound	Upper bound	
λ_{11}	1	–	–	0.74
λ_{21}	0.85	0.11	8.03	0.70
λ_{31}	1.78	0.18	9.25	0.86
λ_{42}	1	–	–	0.77
λ_{52}	4.54	0.50	9.04	0.94
φ_{11}	6.78	1.37	4.96	1
φ_{22}	2.18	0.43	5.11	1
φ_{21}	2.78	0.54	5.16	0.72
θ_{11}	5.69	0.84	6.81	0.46
θ_{22}	5.12	0.707	7.252	0.51
θ_{33}	6.78	1.571	4.314	0.26
θ_{44}	1.52	0.255	5.936	0.41
θ_{55}	5.72	3.945	1.451	0.11

parameter estimates are given in Table 1.4. All standardized factor loadings are larger than 0.70, meaning that they are substantially indicative of the common factor on which they load. The correlation between the common factors internalizing and externalizing is significant and quite large, 0.72. The proportion of explained variance is largest in indicator 5 ($1 - 0.11 = 0.89$) and smallest in indicator 2 ($1 - 0.51 = 0.49$). See Appendix D for an annotated OpenMx-script from this example.

In the two examples given in this chapter the input matrix was a covariance matrix. Maximum likelihood estimation assumes analysis of the covariance matrix, and not of the correlation matrix. However, sometimes only the correlation matrix is available. Treating the correlation matrix as a covariance matrix leads to incorrect results when estimating confidence intervals or when testing specific hypotheses (Cudeck 1989). To obtain correct results, a so-called estimation constraint can be added. This constraint enforces the diagonal of the model implied correlation matrix to always consist of 1's during the estimation.

The factor model and path model are the two basic models within the structural equation modeling framework. Once a factor model has been established, the analysis often goes some steps further, for example by including predictor variables like age to investigate age differences in the latent variables Internalizing and Externalizing problems. Another extension is multigroup modeling, in which a model is fitted to covariance matrices from different groups of respondents simultaneously, giving the opportunity to test the equality of parameters across groups. For example, in the path model from Fig. 1.1, it may be hypothesized that the effect of Positive and Negative relations on Engagement may be stronger for children in elementary school than for children in secondary school.

Some cautions about SEM have to be considered. If a model fits the data well, and is accepted by the researcher as the final model, it does not mean that the model is the correct model in the population. If the model is not rejected, this could be due to lack of statistical power to reject the model. Moreover, there may be other models that fit the data just as well as the hypothesized model. Therefore, it is important to consider equivalent models (MacCallum et al. 1993). If a model is rejected however, the conclusion is that the model does not hold in the population. This chapter is far too short to discuss all relevant issues in SEM. Several books have been written that can be used to learn about SEM, see for example Bollen (1989), Byrne (e.g. 1998), Geiser (2012), Loehlin (1998), and Kline (2011).

1.3 Why Should You Combine SEM and MA?

Most research questions are about relations (or differences) between a set of variables. The hypothetical model in Fig. 1.1 for example, states that positive and negative relations lead to achievement through engagement. Current practice in meta-analysis is to meta-analyze each effect in this model separately. The questions these analyses answer are: What is the pooled effect of positive relations on engagement? And: What is the pooled effect of engagement on achievement? However, what the researcher also may want to know is: Is this model a good representation of the data? Are the effects of positive and negative relations on achievement fully mediated by engagement? Which effects are lacking in this model?

Using MASEM, information from multiple studies is used to test a single model that explains the relationships between a set of variables or to compare several models that are supported by different studies or theories (Becker 1992; Viswesvaran and Ones 1995). MASEM provides the researcher measures of overall fit of a model, as well as parameter estimates with confidence intervals and standard errors. By combining meta-analysis and SEM, some of the difficulties in the separate fields may be overcome.

Structural equation modelling requires large sample sizes. Small samples lead to low statistical power, and non-rejection of models. If several (small) studies investigate the same phenomenon, they may end up with very different final models, leading to a wide array of models describing the same phenomena. By combining the information from several (possibly underpowered) primary studies, general conclusions can be reached about which model is most appropriate. Norton et al. (2013) for example, used MASEM to investigate the factor structure of an anxiety and depression scale, by comparing ten different models that were proposed based on different primary studies. Furthermore, MASEM can be used to answer research questions that are not addressed in any of the primary studies. Even about models that include a set of variables that none of the primary studies included all in their study. For example, Study 1 may report correlations between variable A and variable B. Study 2 may report correlations between variables B and C, and Study 3 between variable A and C. Although none of the studies included all variables, one model can be fit on these three variables using MASEM (Viswesvaran and Ones 1995).

I use the term MASEM for the process of fitting a structural equation model on the combined data from several studies. SEM can also be used to perform ordinary meta-analysis (SEM-based meta-analysis). SEM-based meta-analysis is outside the scope of this book, but see Cheung (2008, 2015) for an explanation.

MASEM is a fairly young field of research, and it seems to be growing in popularity, both in substantive and methodological research. At this moment, a special issue about MASEM is being edited for the journal *Synthesis Research Methods*.

References

Becker, B. J. (1992). Using results from replicated studies to estimate linear models. *Journal of Educational Statistics, 17*, 341–362.

Bentler, P. M. (1990). Comparative fit indexes in structural models. *Psychological Bulletin, 107*, 238–246.

Bollen, K. A. (1989). *Structural equations with latent variables*. New York, NY: Wiley.

Borenstein, M., Hedges, L. V., Higgins, J. P. T., & Rothstein, H. (2009). *Introduction to meta-analysis*. Hoboken, NJ: Wiley.

Browne, M. W., & Cudeck, R. (1992). Alternative ways of assessing model fit. *Sociological Methods & Research, 21*, 230–258.

Byrne, B. M. (1998). *Structural equation modeling with LISREL, PRELIS, and SIMPLIS: Basic concepts, applications, and programming*. Mahwah, NJ: Lawrence Erlbaum Associates Inc.

Card, N. A. (2012). *Applied meta-analysis for social science research*. New York, NY: Guilford Press.

Chalmers, I., Hedges, L. V., & Cooper, H. (2002). A brief history of research synthesis. *Evaluation and the Health Professions, 25*(1), 12–37.

Cheung, M. W.-L. (2008). A model for integrating fixed-, random-, and mixed-effects meta-analyses into structural equation modeling. *Psychological Methods, 13*(3), 182–202.

Cheung, M. W.-L. (2015). *Meta-analysis: A structural equation modeling approach*. Chichester, UK: Wiley.

Cooper, H., Hedges, L. V., & Valentine, J. C. (Eds.). (2009). *The handbook of research synthesis and meta-analysis*. New York, NY: Russell Sage.

Cooper, H. M., & Hedges, L. V. (2009). Research synthesis as a scientific enterprise. In H. Cooper, L. V. Hedges, & J. C. Valentine (Eds.), *The handbook of research synthesis and meta-analysis* (pp. 3–16). New York, NY: Russell Sage.

Cudeck, R. (1989). Analysis of correlation matrices using covariance structure models. *Psychological Bulletin, 105*(2), 317–327.

Geiser, C. (2012). *Data analysis with Mplus*. New York, NY: Guilford.

Glass, G. V. (1976). Primary, secondary, and meta-analysis of research. *The Educational Researcher, 10*, 3–8.

Hedges, L. V., & Vevea, J. L. (1998). Fixed-and random-effects models in meta-analysis. *Psychological Methods, 3*, 486.

Jöreskog, K. G. (1973). Analysis of covariance structures. *Multivariate analysis, 3*, 263–285.

Kline, R. B. (2011). *Principles and practice of structural equation modeling*. New York, NY: Guilford.

Laird, N. M., & Mosteller, F. (1990). Some statistical methods for combining experimental results. *International Journal of Technology Assessment in Health Care, 6*, 5–30.

Lipsey, M. W., & Wilson, D. B. (2001). *Practical meta-analysis.* Thousand Oaks, CA: Sage Publications.

Loehlin, J. C. (1998). *Latent variable models: An introduction to factor, path, and structural analysis.* Mahwah, NJ: Lawrence Erlbaum Associates Publishers.

MacCallum, R. C., Wegener, D. T., Uchino, B. N., & Fabrigar, L. R. (1993). The problem of equivalent models in applications of covariance structure analysis. *Psychological Bulletin, 114*(1), 185–199.

Norton, S., Cosco, T., Doyle, F., Done, J., & Sacker, A. (2013). The Hospital Anxiety and Depression Scale: A meta confirmatory factor analysis. *Journal of Psychosomatic Research, 74*, 74–81.

O'Rourke, K. (2007). An historical perspective on meta-analysis: Dealing quantitatively with varying study results. *Journal of the Royal Society of Medicine, 100*(12), 579–582.

Rothstein, H. R., Sutton, A. J., & Borenstein, M. (Eds.). (2005). *Publication bias in meta-analysis: Prevention, assessment and adjustments.* Hoboken, NJ: Wiley.

Rosenthal, R., & Rubin, D. B. (1978). Interpersonal expectancy effects: The first 345 studies. *Behavioral and Brain Sciences, 1*(03), 377–386.

Rosenthal, R., & Rubin, D. B. (1982). Comparing effect sizes of independent studies. *Psychological Bulletin, 92*(2), 500–504.

Schermelleh-Engel, K., Moosbrugger, H., & Müller, H. (2003). Evaluating the fit of structural equation models: Tests of significance and descriptive goodness-of-fit measures. *Methods of psychological research online, 8*(2), 23–74.

Schmidt, F. L., & Hunter, J. E. (1977). Development of a general solution to the problem of validity generalization. *Journal of Applied Psychology, 62*(5), 529–540.

Sharpe, D. (1997). Of apples and oranges, file drawers and garbage: Why validity issues in meta-analysis will not go away. *Clinical Psychology Review, 17*(8), 881–901.

Smith, M. L., & Glass, G. V. (1977). Meta-analysis of psychotherapy outcome studies. *American Psychologist, 32*, 752–760.

Spearman, C. (1904). "General Intelligence," Objectively Determined and Measured. *The American Journal of Psychology, 15*(2), 201–292.

Spearman, C. (1928). The sub-structure of the mind. *British Journal of Psychology. General Section, 18*(3), 249–261.

Steiger, J. H., & Lind, J. C. (1980). *Statistically based tests for the number of common factors.* Paper presented at the annual meeting of the Psychometric Society, Iowa City, IA (Vol. 758).

Valentine, J. C. (2009). Judging the quality of primary research. In H. Cooper, L. V. Hedges, & J. C. Valentine (Eds.), *The handbook of research synthesis and meta-analysis* (pp. 129–146). New York, NY: Russell Sage.

van Assen, M. A. L. M., van Aert, R. C. M., & Wicherts, J. M. (2014). Meta-analysis using effect size distributions of only statistically significant studies. *Psychological methods.* Advance online publication. http://dx.doi.org/10.1037/met0000025

Viswesvaran, C., & Ones, D. (1995). Theory testing: Combining psychometric meta-analysis and structural equations modeling. *Personnel Psychology, 48*, 865–885.

Wright, S. (1920). The relative importance of heredity and environment in determining the piebald pattern of guinea–pigs. *Proceedings of the National Academy of Sciences, 6*, 320–332.

Chapter 2
Methods for Meta-Analytic Structural Equation Modeling

Abstract The process of performing meta-analytic structural equation modeling (MASEM) consists of two stages. First, correlation coefficients that have been gathered from studies have to be combined to obtain a pooled correlation matrix of the variables of interest. Next, a structural equation model can be fitted on this pooled matrix. Several methods are proposed to pool correlation coefficients. In this chapter, the univariate approach, the generalized least squares (GLS) approach, and the Two Stage SEM approach are introduced. The univariate approaches do not take into account that the correlation coefficients may be correlated within studies. The GLS approach has the limitation that the Stage 2 model has to be a regression model. Of the available approaches, the Two Stage SEM approach is favoured for its flexibility and good statistical performance in comparison with the other approaches.

Keywords Meta-analytic structural equation modeling · Univariate meta-analysis · Multivariate meta-analysis · GLS-approach · Two-stage structural equation modeling · MASEM

2.1 Introduction

As shown in Chap. 1, a structural equation model can be fitted to the covariance or correlation matrix of the variables of interest, without requirement of the raw data. Therefore, if articles report the correlations between the research variables (or information that can be used to estimate the correlation), the results can be used in a meta-analysis. MASEM combines structural equation modeling with meta-analysis by fitting a structural equation model on a meta-analyzed covariance or correlation matrix. As the primary studies in a meta-analysis often involve variables that are measured in different scales, MASEM is commonly conducted using a pooled correlation rather than covariance matrix. In the remainder of this book I will therefore focus on correlation matrices (but see Beretvas and Furlow 2006;

© The Author(s) 2015
S. Jak, *Meta-Analytic Structural Equation Modelling*,
SpringerBriefs in Research Synthesis and Meta-Analysis,
DOI 10.1007/978-3-319-27174-3_2

Cheung and Chan 2009). MASEM typically consists of two stages (Viswesvaran and Ones 1995). In the first stage, correlation coefficients are tested for homogeneity across studies and combined together to form a pooled correlation matrix. In the second stage, a structural equation model is fitted to the pooled correlation matrix. In the next sections I outline the different approaches to pool correlation coefficients under the assumption that the correlations are homogenous across studies (fixed effects approaches). Heterogeneity of correlation coefficients and random effects approaches are discussed in Chap. 3.

2.2 Univariate Methods

In the univariate approaches, the correlation coefficients are pooled separately across studies based on bivariate information only. Dependency of correlation coefficients within studies is not taken into account (as opposed to multivariate methods, described in the next section). In the univariate approaches, a population value is estimated for each correlation coefficient separately. For one correlation coefficient, for each study i, the correlation coefficient is weighted by the inverse of the estimated sampling variance (the squared standard error), v_i. The sampling variance of the correlation between variables A and B is given by:

$$v_{i\_AB} = (1 - \rho_{i\_AB}^2)^2 / n_i, \tag{2.1}$$

where n_i is the sample size in study i, and the observed correlation r_{i_AB} can be plugged in for the unknown population correlation ρ_{i_AB}. By taking the average of the weighted correlation coefficients across the k studies, one obtains the synthesized population correlation estimate:

$$\hat{\rho} = \frac{\sum_{i=1}^{k} \frac{1}{v_{i\_AB}} r_{i\_AB}}{\sum_{i=1}^{k} \frac{1}{v_{i\_AB}}}. \tag{2.2}$$

Weighting by the inverse sampling variance ensures that more weight is given to studies with larger sample size (and thus smaller sampling variance). Because the sampling variance of a correlation coefficient depends on the absolute value of the correlation coefficient, some researchers (e.g. Hedges and Olkin 1985) proposed to use Fisher's z-transformation on the correlation coefficients before synthesizing the values. The estimated sampling variance v_i of a transformed correlation z in a study i is equal to $1/(n_i - 3)$, where n_i is the sample size in study i. After obtaining the pooled z-value, it can be back-transformed to an r-value for interpretation.

There is no consensus on whether it is better to use the untransformed correlation coefficient r or the transformed coefficient z in meta-analysis (see Corey et al. 1998). Hunter and Schmidt (1990) argued that averaging r leads to better estimates of the population coefficient than averaging z. However, several simulation studies (Cheung and Chan 2005; Furlow and Beretvas 2005; Hafdahl and Williams 2009) showed that differences between the two methods were generally very small, but when differences are present, the z approach tends to do better. If a random effects model is assumed however, Schulze (2004) recommends r over z.

If the correlation coefficients are pooled across studies (using the r or z method), one pooled correlation matrix can be constructed from the separate coefficients. The hypothesized structural model is then fit to this matrix, as if it was an observed matrix in a sample.

Apart from the problem that the correlations are treated as independent from each other within a study, the univariate methods have more issues (Cheung and Chan 2005). Because not all studies may include all variables, some Stage 1 correlation coefficients will be based on more studies than others. This leads to several problems. First, it may lead to non-positive definite correlation matrices (Wothke 1993), as different elements of the matrix are based on different samples. Non-positive definite matrices cannot be analysed with structural equation modeling. Second, correlation coefficients that are based on less studies are estimated with less precision and should get less weight in the analysis, which is ignored in the standard approaches. Third, if different sample sizes are associated with different correlation coefficients, it is not clear which sample size has to be used in Stage 2. One could for example use the mean sample size, the median sample size or the total sample size, leading to different results regarding fit indices and statistical tests in Stage 2. Due to these difficulties, univariate methods are not recommended for MASEM (Becker 2000; Cheung and Chan 2005).

2.3 Multivariate Methods

The two best known multivariate methods for meta-analysis are the generalized least squares (GLS) method (Becker 1992, 1995, 2009) and the Two-Stage SEM method (Cheung and Chan 2005). Both will be explained in the next sections.

2.3.1 The GLS Method

Becker (1992, 1995, 2009) proposed using generalized least squares estimation to pool correlation matrices, taking the dependencies between correlations into account. This means that not only the sampling variances in each study are used to weight the correlation coefficients, but also the sampling covariances. The estimate of the population variance of a correlation coefficient was given in Eq. (2.1). The population covariance between two correlation coefficients, let's say between variables A and B and between the variables C and D, is given by the long expression:

$$
\begin{aligned}
\mathrm{cov}\,(\rho_{i\_AB}, \rho_{i\_CD}) = {} & (0.5\rho_{i\_AB}\rho_{i\_BC}(\rho_{i\_AC}{}^2 + \rho_{i\_AD}{}^2 + \rho_{i\_BC}{}^2 + \rho_{i\_BD}{}^2) \\
& + \rho_{i\_AC}\rho_{i\_BD} + \rho_{i\_AD}\rho_{i\_BC} - (\rho_{i\_AB}\rho_{i\_AC}\rho_{i\_AD} + \rho_{i\_AB}\rho_{i\_BC}\rho_{i\_BD} \\
& + \rho_{i\_AC}\rho_{i\_BC}\rho_{i\_CD} + \rho_{i\_AD}\rho_{i\_BD}\rho_{i\_CD}))/n_i,
\end{aligned}
$$

$$(2.3)$$

where ρ_i indicates a population correlation value in study i and n_i is the sample size in study i (Olkin and Siotani 1976). As the population parameters ρ_i are unknown, the estimates of the covariances between correlations can be obtained by plugging in sample correlations for the corresponding ρ_i's in Eq. (2.3). However, because the estimate from a single study is not very stable, it is recommended to use pooled estimates of ρ, by using the (weighted) mean correlation across samples (Becker and Fahrbach 1994; Cheung 2000; Furlow and Beretvas 2005). These pooled estimates should then also be used to obtain the variances of the correlation coefficients (by plugging in the pooled estimate in Eq. 2.1). This way, a covariance matrix of the correlation coefficients, denoted $\mathbf{V}_i$ is available for each study in the meta-analysis. The dimensions of $\mathbf{V}_i$ may differ across studies. If a study includes three variables, and reports the three correlations between the variables, $\mathbf{V}_i$ has three rows and three columns. The values of $\mathbf{V}_i$ are treated as known (as opposed to estimated) in the GLS approach. The $\mathbf{V}_i$ matrices for each study are put together in one large matrix, $\mathbf{V}$, which is a block diagonal matrix, with the $\mathbf{V}_i$ matrix for each study on its diagonal:

$$\mathbf{V} = \begin{bmatrix} \mathbf{V}_1 & 0 & 0 & 0 \\ 0 & \mathbf{V}_2 & \cdots & 0 \\ 0 & \vdots & \ddots & \vdots \\ 0 & 0 & \cdots & \mathbf{V}_K \end{bmatrix}.$$

$\mathbf{V}$ is a symmetrical matrix with numbers of rows and columns equal to the total number of observed correlation coefficients across the studies.

For performing the multivariate meta-analysis using the GLS-approach, two more matrices are needed: A vector with the observed correlations in all the studies, $\mathbf{r}$, and a matrix with zeros and ones that is used to indicate which correlation coefficients are present in each study. The vector with the observed correlations in all studies can be created by stacking the observed correlations in each study in a column vector. The length of this vector will be equal to the total number of correlations in all studies. If all k studies included all p variables, $\mathbf{r}$ will be a pk by 1 vector. Most often, not all studies include all research variables, in which case a selection matrix, $\mathbf{X}$, is needed. For a study i, which for example included variables A and B but not C (and thus reports r_{i_AB}, but not r_{i_AC} and r_{i_BC}), a selection matrix is created by constructing a 3 by 3 identity matrix (a matrix with ones on the diagonal and zeros off-diagonal) and removing the row of the missing correlation. In this study the selection matrix will thus look like this:

$$\begin{bmatrix} 1 & 0 & 0 \\ 0 & 1 & 0 \end{bmatrix},$$

and in a study which included all three correlations, the selection matrix will be an identity matrix:

$$\begin{bmatrix} 1 & 0 & 0 \\ 0 & 1 & 0 \\ 0 & 0 & 1 \end{bmatrix}.$$

Doing this for all k studies, leads to k small matrices with three columns and number of rows equal to the number of present correlations. These matrices are then stacked to create matrix $\mathbf{X}$, which has three columns and number of rows equal to the sum of all correlation coefficients across studies. That is, it has the same number of rows as the stacked vector of observed correlations, $\mathbf{r}$. Using matrix algebra with these three matrices, the estimates of the pooled correlation coefficients can be obtained:

$$\widehat{\boldsymbol{\rho}} = (\mathbf{X}^T\mathbf{V}^{-1}\mathbf{X})^{-1}\mathbf{X}^T\mathbf{V}^{-1}\mathbf{r}), \tag{2.4}$$

where $\widehat{\boldsymbol{\rho}}$ is a p-dimensional column vector with the estimates of the population correlation coefficients, as well as the asymptotic covariance matrix of the parameter estimates, $\mathbf{V}_{GLS}$:

$$\mathbf{V}_{GLS} = \left(\mathbf{X}^T\mathbf{V}^{-1}\mathbf{X}\right)^{-1}. \tag{2.5}$$

The only structural model that can be evaluated directly with the GLS method is the regression model. This is done by creating a matrix with the estimated pooled correlations of the independent variables, say $\mathbf{R}_{INDEP}$, and a vector with estimated pooled correlations of the independent variables with the dependent variables, say $\mathbf{R}_{DEP}$, and using the following matrix equation to obtain the vector of regression coefficients $\mathbf{B}$:

$$\mathbf{B} = \mathbf{R}_{INDEP}^{-1}\mathbf{R}_{DEP}. \tag{2.6}$$

This approach is very straightforward (if you use a program to do the matrix algebra), but it is a major limitation that regression models are the only models that can be estimated this way. In order to fit path models or factor models, one has to use a SEM-program and use the pooled correlation coefficients as input to the program. Treating the pooled correlation matrix as an observed matrix shares problems with the univariate methods, it is unclear which sample size has to be used, and potential differences in precision of correlation coefficients is not taken into account. An alternative way to fit a structural equation model on the pooled correlation matrix is to use the $\mathbf{V}_{GLS}$ matrix as a weight matrix in WLS estimation, similar to the TSSEM, which is explained in the next section. For a detailed and accessible description of the GLS method see Becker (1992) and Card (2012).

2.3.2 Two Stage Structural Equation Modeling (TSSEM)

The TSSEM method was proposed by Cheung and Chan (2005). With TSSEM, multigroup structural equation modeling is used to pool the correlation coefficients at Stage 1. In Stage 2, the structural model is fitted to the pooled correlation matrix, using weighted least squares (WLS) estimation. The weight matrix in the WLS procedure is the inversed matrix with asymptotic variances and covariances of the pooled correlation coefficients from Stage 1. This ensures that correlation coefficients that are estimated with more precision (based on more studies) in Stage 1 get more weight in the estimation of model parameters in Stage 2. The precision of a Stage 1 estimate depends on the number and the size of the studies that reported the specific correlation coefficient.

Stage 1: Pooling correlation matrices Let R_i be the $p_i \times p_i$ sample correlation matrix and p_i be the number of observed variables in the ith study. Not all studies necessarily include all variables. For example, in a meta-analysis of three variables A, B and C, the correlation matrices for the first three studies may look like this:

$$\mathbf{R}_1 = \begin{bmatrix} 1 & & \\ r_{1\_AB} & 1 & \\ r_{1\_AC} & r_{1\_BC} & 1 \end{bmatrix}, \quad \mathbf{R}_2 = \begin{bmatrix} 1 & \\ r_{2\_AB} & 1 \end{bmatrix}, \text{ and } \quad \mathbf{R}_3 = \begin{bmatrix} 1 & \\ r_{3\_BC} & 1 \end{bmatrix}.$$

Here, Study 1 contains all variables, Study 2 misses Variable C, and Study 3 misses Variable A. Similar to the GLS approach, selection matrices are needed to indicate which study included which correlation coefficients. Note however, that in TSSEM, the selection matrices filter out missing *variables* as opposed to missing *correlations* in the GLS-approach, and is thus less flexible in handling missing correlation coefficients (see Chap. 4).

In TSSEM the selection matrices are not stacked into one large matrix. For the three mentioned studies, the selection matrices are identity matrices with the rows of missing variables excluded:

$$\mathbf{X}_1 = \begin{bmatrix} 1 & 0 & 0 \\ 0 & 1 & 0 \\ 0 & 0 & 1 \end{bmatrix}, \quad \mathbf{X}_2 = \begin{bmatrix} 1 & 0 & 0 \\ 0 & 1 & 0 \end{bmatrix}, \text{ and } \quad \mathbf{X}_3 = \begin{bmatrix} 0 & 1 & 0 \\ 0 & 0 & 1 \end{bmatrix}.$$

Next, multigroup structural equation modelling is used to estimate the population correlation matrix $\mathbf{R}$ of all p variables (p is three in the example above). Each study is then viewed as a group. The model for each group i (study) is:

$$\boldsymbol{\Sigma}_i = \mathbf{D}_i \left(\mathbf{X}_i \mathbf{R} \mathbf{X}_i^{\mathsf{T}} \right) \mathbf{D}_i. \tag{2.7}$$

In this model, $\mathbf{R}$ is the $p \times p$ population correlation matrix with fixed 1's on its diagonal, matrix $\mathbf{X}_i$ is the $p_i \times p$ selection matrix that accommodates smaller correlation matrices from studies with missing variables ($p_i < p$), and $\mathbf{D}_i$ is a $p_i \times p_i$ diagonal matrix that accounts for differences in scaling of the variables across the studies. Correct parameter estimates can be obtained using maximum likelihood estimation, optimizing the sum of the likelihood functions in all the studies:

$$\mathrm{F}_{\mathrm{ML}} = \sum_{i=1}^{k} \frac{N_i}{N} \mathrm{F}_{\mathrm{ML}_i}, \tag{2.8}$$

where N_i is the sample size in study i, $N = N_1 + N_2 + \cdots + N_k$, and with F_{MLi} for each study as given in Eq. (1.3). Describing the model in Eq. (2.7) in words, it means that a model is fitted to the correlation matrices of all studies, with the restriction that the population correlations are equal across studies. The diagonal $\mathbf{D}_i$ matrices do not have a particular meaning, other than that they reflect differences in variances across the studies. They are needed because the diagonal of $\mathbf{R}$ is fixed at 1, but the diagonals of $\mathbf{\Sigma}_i$ don't necessarily have to equal 1 due to differences in sample variances.[1] Fitting the model from Eq. (2.7) with a SEM program leads to estimates of the population correlation coefficients, as well as the associated asymptotic variance covariance matrix.

A chi-square measure of fit for the model in Eq. (2.7) is available by comparing its minimum F_{ML} value with the minimum F_{ML} value of a saturated model that is obtained by relaxing the restriction that all correlation coefficients are equal across studies. If a separate $\mathbf{R}_i$ is estimated for each study, the selection matrices $\mathbf{X}_i$ are not needed anymore. The model for a specific study then is:

$$\mathbf{\Sigma}_i = \mathbf{D}_i \mathbf{R}_i \mathbf{D}_i. \tag{2.9}$$

The difference between the resulting minimum F_{ML} values of the models in Eqs. (2.9) and (2.7), multiplied by the total sample size minus the number of studies, has a chi-square distribution with degrees of freedom equal to the difference in numbers of free parameters. If the chi-square value of this likelihood ratio test is significant then the hypothesis of homogeneity must be rejected (see Chap. 3), and the fixed effects Stage 2 model should not be fitted to the pooled Stage 1 matrix. In the remainder of this chapter we assume that homogeneity holds.

Stage 2: Fitting structural equation models Cheung and Chan (2005) proposed to use WLS estimation to fit structural equation models to the pooled correlation matrix $\mathbf{R}$ that is estimated in Stage 1. Fitting the Stage 1 model provides estimates of the population correlation coefficients in $\mathbf{R}$ as well as the asymptotic variances and covariances of these estimates, $\mathbf{V}$. In Stage 2, hypothesized structural equation models can be fitted to $\mathbf{R}$ by minimizing the weighted least squares fit function (also known as the asymptotically distribution free fit function; Browne 1984):

$$F_{WLS} = (\mathbf{r} - \mathbf{r}_{MODEL})^T \mathbf{V}^{-1} (\mathbf{r} - \mathbf{r}_{MODEL}), \tag{2.10}$$

where $\mathbf{r}$ is a column vector with the unique elements in $\mathbf{R}$, $\mathbf{r}_{MODEL}$ is a column vector with the unique elements in the model implied correlation matrix ($\mathbf{R}_{MODEL}$), and $\mathbf{V}^{-1}$ is the inversed matrix of asymptotic variances and covariances that is used as the weight matrix. For example, in order to fit a factor model with q factors, one would specify $\mathbf{R}_{MODEL}$ as

$$\mathbf{R}_{MODEL} = \mathbf{\Lambda} \mathbf{\Phi} \mathbf{\Lambda}^T + \mathbf{\Theta}, \tag{2.11}$$

---

[1] I put an example of an analysis with two groups (studies) on my website (http://suzannejak.nl/masem) to illustrate the function of the $\mathbf{D}$-matrices.

where $\boldsymbol{\Phi}$ is a q by q covariance matrix of common factors, $\boldsymbol{\Theta}$ is a p by p (diagonal) matrix with residual variances, and $\boldsymbol{\Lambda}$ is a p by q matrix with factor loadings. Minimizing the WLS function leads to correct parameter estimates with appropriate standard errors and a WLS based chi-square test statistic T_{WLS} (Cheung and Chan 2005; Oort and Jak 2015).

One can also use the pooled correlation matrix and asymptotic covariance matrix from the GLS approach to fit the Stage 2 model with WLS estimation. Cheung and Chan (2005) compared the TSSEM method with the GLS method and the univariate methods. The GLS method in their study was based on Eq. (2.3), so they used the individual study correlation coefficients and not the pooled correlation coefficients as recommended by Becker and Fahrbach (1994) to calculate the sampling weights. The simulation research showed that the GLS method rejects homogeneity of correlation matrices too often and leads to biased parameter estimates at Stage 2. The univariate methods lead to inflated Type 1 errors, while the TSSEM method leads to unbiased parameter estimates and false positive rates close to the expected rates. The statistical power to reject an underspecified factor model was extremely high for all four methods. The TSSEM method overall came out as best out of these methods. Software to apply TSSEM is readily available in the R-Package metaSEM (Cheung 2015), which relies on the OpenMx package (Boker et al. 2011). This package can also be used for the GLS approach and the univariate approaches. More information about the software that can be used to perform MASEM can be found in Chap. 4.

References

Becker, B. J. (1992). Using results from replicated studies to estimate linear models. *Journal of Educational Statistics, 17*, 341–362.

Becker, B. J. (1995). Corrections to "Using results from replicated studies to estimate linear models". *Journal of Educational Statistics, 20*, 100–102.

Becker, B. J. (2000). Multivariate meta-analysis. In H. E. A. Tinsley & S. D. Brown (Eds.), *Handbook of applied multivariate statistics and mathematical modeling* (pp. 499–525). San Diego: Academic Press.

Becker, B. J. (2009). Model-based meta-analysis. In H. Cooper, L. V. Hedges, & J. C. Valentine (Eds.), *The Handbook of Research Synthesis and Meta-analysis (pp. 377–398)* (2nd ed.). New York: Russell Sage Foundation.

Becker, B. J., Fahrbach, K. (1994). A comparison of approaches to the synthesis of correlation matrices. In annual meeting of the American Educational Research Association, New Orleans, LA.

Beretvas, S. N., & Furlow, C. F. (2006). Evaluation of an approximate method for synthesizing covariance matrices for use in meta-analytic SEM. *Structural Equation Modeling, 13*, 153–185.

Boker, S., Neale, M., Maes, H., Wilde, M., Spiegel, M., Brick, T., et al. (2011). OpenMx: An open source extended structural equation modeling framework. *Psychometrika, 76*, 306–317.

Browne, M. W. (1984). Asymptotically distribution-free methods for the analysis of covariance structures. *British Journal of Mathematical and Statistical Psychology, 37*(1), 62–83.

Card, N. A. (2012). *Applied meta-analysis for social science research*. New York: Guilford.

Cheung, M. W.-L. (2015). MetaSEM: An R package for meta-analysis using structural equation modeling. *Frontiers in Psychology*, 5(1521). doi:10.3389/fpsyg.2014.01521

Cheung, M. W.-L., & Chan, W. (2005). Meta-analytic structural equation modeling: A two-stage approach. *Psychological Methods, 10*, 40–64.

Cheung, M. W.-L., & Chan, W. (2009). A two-stage approach to synthesizing covariance matrices in meta-analytic structural equation modeling. *Structural Equation Modeling: A Multidisciplinary Journal, 16*(1), 28–53.

Cheung, S. F. (2000). *Examining solutions to two practical issues in meta-analysis: Dependent correlations and missing data in correlation matrices.* Unpublished doctoral dissertation, Chinese University of Hong Kong.

Corey, D. M., Dunlap, W. P., & Burke, M. J. (1998). Averaging correlations: Expected values and bias in combined Pearson rs and Fisher's z transformations. *The Journal of General Psychology, 125*(3), 245–261.

Furlow, C. F., & Beretvas, S. N. (2005). Meta-analytic methods of pooling correlation matrices for structural equation modeling under different patterns of missing data. *Psychological Methods, 10*(2), 227–254.

Hafdahl, A. R., & Williams, M. A. (2009). Meta-analysis of correlations revisited: Attempted replication and extension of Field's (2001) simulation studies. *Psychological Methods, 14*(1), 24–42. doi10.1037/a0014697

Hedges, L. V., & Olkin, I. (1985). *Statistical models for meta-analysis.* New York, NY: Academic Press.

Hunter, J. E., & Schmidt, F. L. (1990). *Methods of meta-analysis. Correcting error and bias in research findings.* Newbury Park, CA: Sage Publications.

Olkin, I., & Siotani, M. (1976). Asymptotic distribution of functions of a correlation matrix. In S. Ikeda, et al. (Eds.), *Essays in probability and statistics.* Tokyo: Shinko Tsusho Co., Ltd.

Oort, F. J. & Jak, S. (2015). Maximum likelihood estimation in meta-analytic structural equation modeling. doi:10.6084/m9.figshare.1292826

Schulze, R. (2004). *Meta-analysis: A comparison of approaches.* Toronto: Hogrefe & Huber Publishers.

Viswesvaran, C., & Ones, D. S. (1995). Theory testing: Combining psychometric meta-analysis and structural equations modeling. *Personnel Psychology, 48*(4), 865–885.

Wothke, W. (1993). Nonpositive definite matrices in structural modeling. In K. A. Bollen & J. S. Long (Eds.), *Testing structural equation models* (pp. 256–293). Newbury Park, CA: Sage.

Chapter 3
Heterogeneity

Abstract Fixed effects models assume that all differences between correlation coefficients are due to sampling fluctuations, and do not allow inference beyond the studies included in the meta-analysis. Random effects models are more appropriate when researchers wish to make more general statements. Differences between studies' coefficients may occur for other reasons than sampling, for example because other measurement instruments were used or because characteristics of the samples are different. Random effects meta-analytic structural equation modeling takes the study level variance into account. This chapter shows how one can test for heterogeneity of correlation coefficients, and how to quantify the size of the heterogeneity. If heterogeneity is present, the fixed effects model is not appropriate. One option is to explain all heterogeneity with study level variables, for example using subgroup analysis. Random effects analysis can also be combined with subgroup analysis, by fitting a random effects model to subgroups of studies.

Keywords Meta-analytic structural equation modeling · Heterogeneity · Q-test · I^2 · Random effects model · Subgroup analysis · Mixed effects model · MASEM

3.1 Introduction

The univariate and multivariate approaches outlined in the previous chapter are based on the fixed effects model. This means that they assume that there is one true value of the underlying population parameter (correlation coefficient) and all differences in the estimates between studies are due to sampling fluctuations. The goal of the Stage 1 analysis is to estimate the true population value of the correlation coefficient. In the random effects model, it is not assumed that each study has the same underlying population parameter. Instead, each study has its own population correlation coefficient. The goal of the analysis is not to estimate

© The Author(s) 2015
S. Jak, *Meta-Analytic Structural Equation Modelling*,
SpringerBriefs in Research Synthesis and Meta-Analysis,
DOI 10.1007/978-3-319-27174-3_3

one true population value, but the mean and variance of the distribution of population values in all the studies. The distribution of the population values is commonly assumed to be normal. The choice between one of the two approaches is most often based on the differences in interpretation between the two approaches. Results from a fixed effects meta-analysis cannot be generalized to studies that were not included in the analysis, while results from a random effects analysis can (Hedges and Vevea 1998). Random effects analysis may thus often be the most appropriate method for researchers who wish to make general statements. Random effects models take study heterogeneity (differences due to other sources than sampling fluctuations) into account. It may be informative to test whether heterogeneity is present and how large the heterogeneity is.

3.2 Testing the Significance of Heterogeneity

Under a random effects model, the observed effect size (correlation coefficient in our case) in study i can be decomposed in three parts:

$$r_i = \rho_R + u_i + \varepsilon_i, \tag{3.1}$$

where ρ_R indicates the mean of the distribution of correlation coefficients, u_i is the deviation of study i's population correlation coefficient from the average correlation coefficient, and ε_i is the sampling deviation of study i from its study specific population correlation coefficient. If u_i is zero for all studies, the random effects model is equivalent to the fixed effects model. If u_i is not zero for all studies, its variance gives an idea how much heterogeneity there is. The variance of u_i is often denoted with τ^2. The variance of ε_i is the sampling variance v_i, as described in Eq. (2.1).

Whether correlation coefficients can be considered homogenous across studies (whether $\tau^2 = 0$) is usually tested using the Q-test (Cochran 1954). Viechtbauer (2007) found that for the raw correlation coefficient, Type 1 error of the Q-statistic was highly inflated. Therefore, it is recommended to perform the Q-test with the Fisher transformed correlation coefficient, z_i. Other tests than the Q-test exist (see Viechtbauer 2007), but for the Fisher transformed correlation coefficient, the Q-test is shown to keep the best control of Type 1 errors, given that the sample sizes of the included studies are large enough. The Q-statistic for a specific transformed correlation coefficient z_i is calculated as:

$$Q = \sum_{i=1}^{k} (w_i(z_i - \hat{\rho})^2), \tag{3.2}$$

where w_i is $1/v_i$, z_i is the transformed effect size in study i, and $\hat{\rho}$ is the weighted average effect size (see Eq. 2.2). When homogeneity holds, Q approximately follows a chi-square distribution with degrees of freedom equal to the number of studies k minus 1. So, the calculated Q-value may be compared to the critical chi-square value given degrees of freedom and alpha, to test the significance of the Q-statistic. If the Q-statistic is significant, the conclusion is that there is significant heterogeneity.

A multivariate version of the Q-test also exists, based on the GLS approach (Becker 1992, 1995; Cheung and Chan 2005a). Using the matrices from paragraph 2.3.1, the Q_{GLS} statistic is:

$$Q_{GLS} = \mathbf{r}^T(\mathbf{V}^{-1} - \mathbf{V}^{-1}\mathbf{X}(\mathbf{X}^T\mathbf{V}^{-1}\mathbf{X})^{-1}\mathbf{X}^T\mathbf{V}^{-1})\mathbf{r}, \tag{3.3}$$

which, if homogeneity holds, theoretically follows a chi-square distribution with degrees of freedom equal to the total number of the observed correlation coefficients in all studies, minus the number of population correlation coefficients to be estimated. Simulations by Cheung and Chan (2005a) and Becker and Fahrbach (1994) showed that the rejection rate of the Q_{GLS} statistic was far above the nominal alpha level, so homogeneity was rejected too often.

Instead of the multivariate Q-test, one could evaluate the univariate Q-tests for all correlation coefficients, using a Bonferroni adjusted alpha level. If one of the correlation coefficients shows significant heterogeneity, the hypothesis of homogeneity of the correlation matrix should be rejected. This approach was proposed by Cheung (2000), and has been found to have acceptable rejection rates.

A more obvious test on the homogeneity of correlation coefficients is based on the fit of the Stage 1 model from the TSSEM approach. Stage 1 involves a comparison of a model in which all correlation coefficients are set equal across studies, with a model in which all correlation coefficients are freely estimated across studies. If the constrained model fits the data significantly worse, homogeneity should be rejected. Because the model in which all correlation coefficients are freely estimated is saturated (has zero degrees of freedom), the overall χ^2-value with the associated degrees of freedom of the constrained model provides a test for homogeneity. This χ^2-test of the TSSEM approach has been found to perform well (Cheung and Chan 2005a).

3.3 The Size of the Heterogeneity

If significant heterogeneity of the correlation coefficients has been found, it may be of interest to quantify the size of the heterogeneity. Higgins and Thompson (2002) proposed three suitable measures, of which the I^2 measure is most used and has the most convenient interpretation. The I^2 of a set of effect sizes in different studies is interpreted as the proportion of the total variability that is due to differences between studies. In a random effects model, the total variance of a specific effect size consist of the variance of u_i, τ^2, and the sampling variance v_i. The I^2 measure is calculated using the "typical" sampling variance (see Higgins and

Thompson), that is assumed to be equal across studies ($v_i = v$), and can be estimated using the Q-statistic as:

$$I^2 = \frac{\tau^2}{\tau^2 + v} = \frac{Q - (k - 1)}{Q}, \tag{3.4}$$

where Q is defined in Eq. (3.2). If Q is lower than expected (lower than the
degrees of freedom $k - 1$), then I^2 is restricted to zero. As can be deduced from
Eq. (3.4), Q can be seen as a measure of the overall heterogeneity. The expected
variability due to sampling fluctuations is equal to $k - 1$. So, I^2 gives the proportion of variability in effect sizes other than sampling variability. I^2 values of 0.25,
0.50 and 0.75 are used as rules of thumb to indicate low, medium and high levels
of heterogeneity (Higgins et al. 2003).

Several other definitions of I^2 have been proposed, using different choices of
the typical sampling variance (see Takkouche et al. 1999, and Xiong et al. 2010).
The I^2 coefficient is most commonly calculated separately for each correlation
coefficient. Multivariate extensions of I^2 are also proposed (Jackson et al. 2012)
but still need more evaluation.

3.4   Random Effects Analysis or Explaining Heterogeneity

If correlation coefficients in MASEM are heterogeneous across studies, two
options are available to handle the heterogeneity. One option is to use random
effects modeling, which means that the between-studies variance is estimated,
and the Stage 1 pooled correlations are estimated as weighed averaged correlation
coefficients, where the weights involve both between-studies and within-studies
(sampling) variance. Another approach is to explain all heterogeneity by study-
level variables. These study-level variables are called moderators, because they
moderate the relations between study variables. If the moderator variables explain
all differences between studies, the residual between-study variance is zero, and a
fixed effects model applies.

3.4.1   Random Effects MASEM

Stage 1 analysis When random effects MASEM is used, the between-study heterogeneity is taken into account by estimating study-level variance of the correlation coefficients in Stage 1. In a random effects model, the correlation matrices
are not only weighted by the sampling variance (v_i), but also by the between-study
variance (τ^2). In univariate analysis it means that a specific correlation coefficient
r_i is weighed with $1/(v_i + \tau^2)$. Because the between-study variance is equal across

all studies, the random effects weight is just the fixed effects weight with a constant added to the denominator. One consequence is that the weights the different studies get are relatively more equal, so small studies get relatively more weight in estimating the average effect size, and very large studies get relatively less weight than in the fixed effects model. Another consequence is that the standard errors and confidence intervals of parameter estimates in a random effects model will be larger, leading to less significant results than the fixed effects model.

With multivariate random effects modelling, the matrix with the weights, $\mathbf{V}$, is adjusted to account for the between-studies variance and covariance. For one study, it means that a matrix with between-study variance and covariance of the correlation coefficients is added to the matrix with sampling variance and covariance. The random effects model for a vector of correlation coefficients for a study i decomposes the vector in three parts:

$$\mathbf{r}_i = \rho_R + \mathbf{u}_i + \varepsilon_i, \tag{3.5}$$

where ρ_R indicates the vector of means of the correlation coefficients, $\mathbf{u}_i$ is a vector of deviations of study i's population correlation coefficients from the average correlation coefficients, and ε_i is a vector with the sampling deviations of study i from its study specific population correlation coefficients. The covariance of $\mathbf{u}_i$ denotes the matrix with study level variance and covariance, $\mathbf{T}^2$. The covariance of ε_i denotes the matrix with sampling variance and covariance in study i, $\mathbf{V}_i$. The weight matrix in the random effects analysis is the sum of $\mathbf{T}^2$ and $\mathbf{V}_i$.

The between-studies variance covariance matrix $\mathbf{T}^2$ can be estimated using different approaches. The method of moments uses an estimator of $\mathbf{T}^2$ that is based on the Q-statistic from a fixed effects model (DerSimonian and Laird 1986). Using this estimator leads to the multilevel version of the GLS-approach (Becker 1992, 1995). In the two-stage approach, random effects TSSEM is performed using maximum likelihood estimation (Cheung 2013), in which ρ_R and $\mathbf{T}^2$ are estimated simultaneously. The random effects TSSEM is presented by Cheung (2014).

Estimating the between-study variances in $\mathbf{T}^2$ is relatively simple, but the estimation of the between-study covariance often gives problems, particularly with small numbers of studies or small heterogeneity. If this is the case, Becker (2009) advises to add the between-studies variances (the diagonal of the covariance matrix) to the weight matrix only.

Stage 2 analysis In random effects TSSEM, fitting the structural model (Stage 2) is very similar to the fixed effects approach. The difference is that now the averaged correlation matrix $\mathbf{R}_R$ from a random effects analysis is used as the input matrix for the structural equation model, and the weight matrix $\mathbf{V}_R$ from a random effects model is used in the WLS-fit function:

$$F_{WLS} = (\mathbf{r}_R - \mathbf{r}_{MODEL})^T \mathbf{V}_R^{-1} (\mathbf{r}_R - \mathbf{r}_{MODEL}), \tag{3.6}$$

where $\mathbf{r}_R$ is a vector with the unique elements of the averaged correlation matrix $\mathbf{R}_R$ from a random effects analysis and $\mathbf{V}_R$ is the asymptotic variance covariance matrix associated with $\mathbf{R_R}$. The between-studies variance does not play a role directly in the Stage 2 model, it is filtered out in the Stage 1 analysis. In the GLS approach, one can obtain the regression coefficients using Eq. (2.6), but with an $\mathbf{R_{INDEP}}$ and $\mathbf{R_{DEP}}$ obtained from a random effects Stage 1 analysis. Alternatively, one can use the pooled correlation and asymptotic covariance matrix from a random effects GLS-analysis in Eq. (3.6).

3.4.2 Subgroup Analysis

Another solution to heterogeneity in correlation matrices is to explain all heterogeneity using study level variables. In MASEM, subgroups of studies are created based on values of the (categorical or categorized) study-level variables (Cheung and Chan 2005b). Grouping variables may for example include the country in which the study is conducted, the age of the respondents in the study and the population under consideration in the study (e.g. patients vs. non-patients). If the moderator variable explains all heterogeneity, the correlation coefficients are homogenous within subgroups. With subgroup analysis, each subgroup has its own pooled correlation matrix at Stage 1, and the structural model is fit independently to the matrices of the subgroups. An advantage of performing subgroup analysis is that the effect of study-level moderators is explicitly tested. A disadvantage is that it may lead to the investigation of many subgroups containing small numbers of studies. Moreover, not all heterogeneity may be explained by the moderators. If the researchers have a substantive interest in the moderators, and do not believe that the moderator should explain all heterogeneity, one can also perform subgroup analysis and fit a random effects model in each subgroup.

The primary reason to perform a subgroup analysis will often be that the researchers have hypotheses about differences between subgroups, and not just to explain away heterogeneity. An example of a MASEM analysis in which subgroup analysis is interesting from a theoretical point of view is performed by Roorda et al. (under review). They investigated the influence of positive and negative teacher-student relations on student engagement and student achievement (see the example from Chap. 1). They expected that the path coefficients would be different across samples from primary schools and samples from secondary schools. Indeed, it appeared that the effect of positive relations on engagement was significantly stronger in samples from secondary schools. Testing the equality of parameters across subgroups is not readily implemented in the metaSEM package, but can be performed by using a SEM program directly to analyse the Stage 2 model with WLS-estimation. Fixed effects and random effects MASEM with subgroup analysis using the metaSEM package will be illustrated in Chap. 5.

References

Becker, B. J. (1992). Using results from replicated studies to estimate linear models. *Journal of Educational Statistics, 17*, 341–362.

Becker, B. J. (1995). Corrections to using results from replicated studies to estimate linear models. *Journal of Educational Statistics, 20*, 100–102.

Becker, B. J. (2009). Model-based meta-analysis. In H. Cooper, L. V. Hedges, & J. C. Valentine (Eds.), *The handbook of research synthesis and meta-analysis* (2nd ed., pp. 377–398). New York (NY): Russell Sage Foundation.

Becker, B. J., & Fahrbach, K. (1994). *A comparison of approaches to the synthesis of correlation matrices*. New Orleans, LA: In annual meeting of the American Educational Research Association.

Cheung, M. W.-L. (2013). Multivariate meta-analysis as structural equation models. *Structural Equation Modeling, 20*(3), 429–454.

Cheung, M. W.-L. (2014). Fixed- and random-effects meta-analytic structural equation modeling: Examples and analyses in R. *Behavior Research Methods, 46*, 29–40.

Cheung, M. W.-L., & Chan, W. (2005a). Meta-analytic structural equation modeling: A two-stage approach. *Psychological Methods, 10*, 40–64.

Cheung, M. W.-L., & Chan, W. (2005b). Classifying correlation matrices into relatively homogeneous subgroups: A cluster analytic approach. *Educational and Psychological Measurement, 65*, 954–979.

Cheung, S. F. (2000). *Examining solutions to two practical issues in meta-analysis: Dependent correlations and missing data in correlation matrices*. Unpublished doctoral dissertation, Chinese University of Hong Kong.

Cochran, W. (1954). The combination of estimates from different experiments. *Biometrics, 10*(1), 101–129.

DerSimonian, R., & Laird, N. (1986). Meta-analysis in clinical trials. *Controlled Clinical Trials, 7*(3), 177–188.

Hedges, L. V., & Vevea, J. L. (1998). Fixed-and random-effects models in meta-analysis. *Psychological Methods, 3*(4), 486–504.

Higgins, J., & Thompson, S. G. (2002). Quantifying heterogeneity in a meta-analysis. *Statistics in Medicine, 21*(11), 1539–1558.

Higgins, J., Thompson, S. G., Deeks, J. J., & Altman, D. G. (2003). Measuring inconsistency in meta-analysis. *British Medical Journal, 327*, 557–560.

Jackson, D., White, I. R., & Riley, R. D. (2012). Quantifying the impact of between-study heterogeneity in multivariate meta-analyses. *Statistics in Medicine, 31*(29), 3805–3820.

Roorda, D. L, Jak, S., Oort, F. J., & Koomen, H. M. Y. (under review). Teacher-student relationships and students' achievement: Using a meta-analytic approach to test the mediating role of school engagement.

Takkouche, B., Cadarso-Suárez, C., & Spiegelman, D. (1999). Evaluation of old and new tests of heterogeneity in epidemiologic meta-analysis. *American Journal of Epidemiology, 150*(2), 206–215.

Viechtbauer, W. (2007). Hypothesis tests for population heterogeneity in meta-analysis. *British Journal of Mathematical and Statistical Psychology, 60*(1), 29–60.

Xiong, C., Miller, J. P., & Morris, J. C. (2010). Measuring study-specific heterogeneity in meta-analysis: application to an antecedent biomarker study of Alzheimer's disease. *Statistics in biopharmaceutical research, 2*(3), 300–309.

Chapter 4
Issues in Meta-Analytic Structural Equation Modeling

Abstract This chapter provides short overviews of unresolved issues in MASEM. The first part of this chapter describes software that can be used to conduct MASEM using TSSEM, the GLS-method and the univariate method. The metaSEM-package is very useful for MASEM. Analyses using this package are shown in the last two chapters of this book. The second issue is about the use of different fit-indices to evaluate the homogeneity of correlation matrices at Stage 1 of TSSEM. The third issue is about handling missing correlations in specific studies. The basic approach is to delete a variable that is associated with a missing correlation, but more efficient methods are possible. The last issue is about a recent adaptation to the existing MASEM approach that may have advantages for handling heterogeneity. The adaptation involves a Stage 2 analysis based on a multi-group model.

Keywords Meta-analytic structural equation modeling · Software · MetaSEM · OpenMx · Fit-indices · Maximum likelihood · Missing correlations

4.1 Software to Conduct MASEM

In principle, all structural equation modeling software can be used to perform meta-analytic structural equation modeling. However, it may involve some complex programming to set up the right model. The easiest way to perform TSSEM is to use the dedicated R-package metaSEM (Cheung 2015a). It requires some basic knowledge of the R-program (see below), but the package itself is quite user friendly. It includes functions to fit the fixed effects Stage 1 model, the random effects Stage 1 model, and to fit the Stage 2 model to the pooled correlation matrix from Stage 1. The package includes several convenient functions to read in the data and to extract parts of the output. It also includes all functions to do standard meta-analysis. Cheung (2015b) gives an overview of the many possibilities with the metaSEM-package.

© The Author(s) 2015
S. Jak, *Meta-Analytic Structural Equation Modelling*,
SpringerBriefs in Research Synthesis and Meta-Analysis,
DOI 10.1007/978-3-319-27174-3_4

Fixed effects MASEM based on the GLS approach can also be performed using the metaSEM-package by constraining the random effects to be zero in the random effects function, but the function uses maximum likelihood estimation. I added an example of the original GLS-approach using R on my website (http://suzannejak.nl/masem).

As the multivariate methods are found to perform better than the univariate methods (see Chap. 2), it is not recommended to perform MASEM using the univariate methods. If one still wants to use them, one could in principle use any meta-analysis program to pool the correlation coefficients in Stage 1, and use any structural equation modeling program to fit the Stage 2 model. In order to pool the correlation coefficients, the R-packages 'metafor' (Viechtbauer 2010) and 'metaSEM' (Cheung 2015a) are very useful. David Wilson (Lipsey and Wilson 2001) has written macros for SPSS, SAS, and STATA to carry out univariate meta-analysis. The macro's are available from his website: (http://mason.gmu.edu/~dwilsonb/ma.html). Several other commercial software programs exist. See Bax et al. (2007) for a comparison of several programs.

For Stage 2 you need a SEM-program. Freely available software packages to conduct structural equation modeling are the R-packages Lavaan (Rosseel 2012) and OpenMx (Boker et al. 2011). In addition there are commercial programs such as Mplus (Muthén and Muthén 2012) and Lisrel (Jöreskog and Sörbom 1996). For the Stage 2 analysis with WLS-estimation, OpenMx and Lisrel are most suitable, as Mplus and Lavaan cannot read in the weight matrix in addition to the pooled correlation matrix.

The freely available programs are packages in R. Therefore, in order to conduct MASEM it is very convenient to be familiar with the R-program. R is a free software environment for statistical computing and graphics. Learning R may be a bit daunting in the beginning, but soon will pay back the effort. To get started with R, several manuals can be found under the contributed documentation on www.r-project.org. For example, these two documents provide a short overview of R (and explain how to install R), and will provide you with enough R-knowledge to be able to use the metaSEM package.

- Marthews, D. (2014). The friendly beginners' R course. http://cran.r-project.org/other-docs.html. Accessed 08 Jan 2015.
- Paradis, E. (2005). R for Beginners. http://cran.r-project.org/other-docs.html. Accessed 08 Jan 2015.

The metaSEM-package uses OpenMx in the background to fit all models. OpenMx is a package in R that can be used for structural equation modeling. OpenMx is very flexible, because the user can use all possibilities of the R-programming environment. This makes OpenMx a suitable program to use in the specification of meta-analytic structural equation models. Because for the MASEM researcher it may be useful to understand OpenMx, I included annotated examples of fitting a path model and a factor model in OpenMx in Appendices B and D.

4.2 Fit-Indices in TSSEM

The chi-square measure of fit can be used in Stage 1 to test the homogeneity of correlation matrices across samples. The chi-square test has as the null hypothesis that the model holds exactly in the population, so all differences between the observed and population matrices are due to sampling. In structural equation modeling it is common to look at measures of approximate or relative fit as well. The Root Mean Squared Error or Approximation (RMSEA, Steiger and Lind 1980) for example, is a measure of approximate fit. The RMSEA is based on the idea that models are approximations to reality and do not have to reflect reality perfectly (MacCallum 2003). If a researcher uses the RMSEA to evaluate the fit of a Stage 1 model in MASEM, he or she implicitly assumes that homogeneity does not have to hold exactly but only approximately. However, it is unclear how much deviation from homogeneity is acceptable when fitting the Stage 2 model under a fixed effects model. At some point, the parameters in the Stage 2 model will become biased and confidence intervals may become too small. Research using simulated data, varying for example the amount and type of heterogeneity (heterogeneity in one or all correlation coefficients), would be needed to evaluate the RMSEA values that are associated with unacceptable heterogeneity.

The CFI is based on a comparison of the fit of the specified model with the fit of the independence model, which is a model in which all variables are assumed to be independent. The CFI strongly depends on the size of the observed correlations. The lower the observed correlations, the better the independence model will fit the data, the lower CFI will be. Because the size of the correlations should not play a role in evaluating heterogeneity, I expect that the CFI is not very useful to evaluate the homogeneity of correlation coefficients in MASEM.

The Standardized Root Mean Squared Residual (SRMSR) is based on the differences between the observed and model implied correlation coefficients. Larger differences between the correlation coefficients will lead to a larger SRMSR, so the SRMSR seems to be useful to evaluate homogeneity at Stage 1. However, just as with the RMSEA, simulation research is needed to evaluate the critical SRMSR values associated with unacceptable heterogeneity.

4.3 Missing Correlations in TSSEM

In fixed effects two-stage SEM, it is no problem when some studies do not include all relevant variables. The missing variables will just be filtered out in the analysis. It is a problem if there are missing correlations for variables that are included in the study. Ideally, researchers always report the correlations between all variables in their study. However, often not all correlations between the research variables are given in a paper. Sometimes, the missing correlations can be derived from other statistics the authors do provide, such as regression coefficients. This is not always possible, for example when two variables are both outcome variables in regression analyses. In the random effects Stage 1 analysis, missing correlations are not a problem, but in the fixed effects analysis they are. As a consequence, for each missing correlation, one of the two variables associated with the correlation has to be treated as missing. Preferably, one would delete the variable with the least remaining correlations with other variables.

Methods to handle missing correlation coefficients in TSSEM more efficiently have been proposed by Jak et al. (2013) and Cheung (2014). Both methods are based on the idea of fixing the missing correlations at some appropriate value (a value that does not lead to a non-positive definite correlation matrix), for example at zero, and estimating an extra parameter for each missing correlation. This way, the fixed values for the missing correlations do not affect the results, and all correlations that are present are used in the analyses. These methods are not implemented in the metaSEM package yet. So, in order to use these methods one will have to specify the needed models in OpenMx directly, or use the program to generate syntax to conduct fixed effects TSSEM with Lisrel (Cheung 2009). A possible problem with this approach is that the fit of the independence model may not be appropriate anymore due to the fixed zeros in the observed correlation matrices (Cheung 2015b). The fit of the independence model is used when calculating some fit-indices, like the CFI. However, the problem of the missing correlations plays a role in Stage 1 of the analysis, and as discussed earlier, the CFI may not be the most appropriate fit measure to evaluate the homogeneity of correlation matrices.

4.4 The ML-Approach to MASEM

A recent alternative to estimating the Stage 2 model in the two-stage approach is to use a maximum likelihood (ML) approach (Oort and Jak 2015). In this approach, multigroup analysis is used for all models. The test of homogeneity of correlation matrices (Stage 1) is identical to TSSEM. The difference lies in fitting the structural model. In the ML-approach, a common $\mathbf{R}_{\text{MODEL}}$ is fitted to the observed matrices or all studies, where $\mathbf{R}_{\text{MODEL}}$ may have the structure of any structural equation model. For example, if one would fit a factor model in Stage 2, the model for each study i would be:

$$\boldsymbol{\Sigma}_i = \mathbf{D}_i\big(\mathbf{X}_i\mathbf{R}_{\text{MODEL}}\mathbf{X}_i^{\mathsf{T}}\big)\mathbf{D}_i,$$
with
$$\mathbf{R}_{\text{MODEL}} = \boldsymbol{\Lambda}\boldsymbol{\Phi}\boldsymbol{\Lambda}' + \boldsymbol{\Theta}. \tag{4.1}$$

Here, $\mathbf{D}_i$ and $\mathbf{X}_i$ are the diagonal and selection matrices defined in Chap. 2, $\boldsymbol{\Lambda}$ is a matrix of factor loadings, $\boldsymbol{\Phi}$ is a matrix with factor variances and covariances, and $\boldsymbol{\Theta}$ is a matrix with residual variances (and covariances). Because $\mathbf{R}_{\text{MODEL}}$ is a restriction of $\mathbf{R}$ in the Stage 1 model, the difference between the associated chi-square values has a chi-square distribution itself with degrees of freedom equal to the difference in the numbers of free parameters in $\mathbf{R}$ and $\mathbf{R}_{\text{MODEL}}$. Oort and

Jak (2015) used simulated data to show that using maximum likelihood estimation in both stages of meta-analysis through SEM leads to almost identical results as using WLS-estimation in Stage 2 of the analysis. The differences in estimation bias, power rate and Type 1 error rates were not consistent and hardly noticeable.

There are some fundamental and practical differences which may guide a researcher's choice between the two methods. Advantages of the ML procedure are that the same estimation method is used at both stages, and that the Stage 1 and Stage 2 models are nested. The ML-procedure may also provide more flexibility in the application of equality constraints across studies in the structural model. In principle, some Stage 2 parameters could be set equal across a subset of studies, another parameter could be set equal across another subset of studies and other parameters could be freely estimated in all studies. Disadvantage of the ML-approach are that it is currently limited to fixed effects models, and that no readily available software package to apply the method exists. The WLS-procedure has practical advantages. In the WLS procedure, the Stage 2 model is not a multi-group model, so that estimation convergence is much faster than in the ML-approach. The necessity to calculate a weight matrix (the inverse of the matrix of asymptotic variances and covariances of the pooled correlation coefficients) may count as a disadvantage of the WLS method, but fortunately the readily available R package metaSEM takes this burden off the user's hands. As a result, the WLS-approach may actually be easier to take than the ML-approach.

References

Bax, L., Yu, L.-M., Ikeda, N., & Moons, K. G. M. (2007). A systematic comparison of software dedicated to meta-analysis of causal studies. *BMC Medical Research Methodology, 7,* 40. doi:10.1186/1471

Boker, S., Neale, M., Maes, H., Wilde, M., Spiegel, M., Brick, T., & Fox, J. (2011). OpenMx: An open source extended structural equation modeling framework. *Psychometrika, 76*(2), 306–317.

Cheung, M. W.-L. (2009). TSSEM: A LISREL syntax generator for two-stage structural equation modeling (Version 1.11) [Computer software and manual].

Cheung, M. W.-L. (2014). Fixed- and random-effects meta-analytic structural equation modeling: Examples and analyses in R. *Behavior Research Methods, 46,* 29–40.

Cheung, M. W.-L. (2015a). MetaSEM: An R package for meta-analysis using structural equation modeling. *Frontiers in Psychology, 5,* 1521. doi:10.3389/fpsyg.2014.01521

Cheung, M. W.-L. (2015b). *Meta-analysis: A structural equation modeling approach.* Chichester, UK: Wiley.

Jak, S., Oort, F. J., Roorda, D. L., & Koomen, H. M. Y. (2013). Meta-analytic structural equation modelling with missing correlations. *Netherlands Journal of Psychology, 67*(4), 132–139.

Jöreskog, K. G., & Sörbom, D. (1996). *LISREL 8: Users' reference guide.* Chicago: Scientific Software International.

Lipsey, M. W., & Wilson, D. B. (2001). *Practical meta-analysis.* Thousand Oaks, CA: Sage.

MacCallum, R. C. (2003). 2001 Presidential address: Working with imperfect models. *Multivariate Behavioral Research, 38,* 113–139. doi:10.1207/S15327906MBR3801_5

Muthén, L. K., & Muthén, B. O. (1998–2012). *Mplus user's guide* (7th ed.). Los Angeles, CA: Muthén & Muthén.

Oort, F. J., & Jak, S. (2015). Maximum likelihood estimation in meta-analytic structural equation modeling. 10.6084/m9.figshare.1292826

Rosseel, Y. (2012). Lavaan: An R package for structural equation modeling. *Journal of Statistical Software, 48*(2), 1–36.

Steiger, J. H., & Lind, J. M. (1980). *Statistically-based tests for the number of common factors.* Paper presented at Psychometric Society Meeting, Iowa City, Iowa.

Viechtbauer, W. (2010). Conducting meta-analyses in R with the metafor package. *Journal of Statistical Software, 36*(3), 1–48.

Chapter 5
Fitting a Path Model with the Two-Stage Approach

Abstract In the first chapter of this book, I presented a path model on four variables involving teacher-student interactions, engagement and achievement at school. This chapter uses data from 45 studies who reported (a subset of) the correlations between these variables and the percentage of students with low socio-economic status in the classroom. Using the metaSEM package, it is illustrated how the data can be prepared for analysis, how to fit a fixed and random effects Stage 1 model to the correlation matrices, and how to specify the hypothesized Stage 2 model. Models are fitted to the overall data and to subgroups with low versus high socio-economic status. All steps that have to be taken to perform the analyses are discussed, as well as the relevant output.

Keywords Meta-analytic structural equation modeling · metaSEM · Path model · Mediation · Teacher-student relations · Engagement · Achievement · SES

5.1 Introduction

Roorda et al. (2011) collected data from 99 studies that reported correlations between positive teacher-student relations, negative teacher-student, student engagement and student achievement. Correlations between positive teacher-student relations and negative teacher-student relations were collected afterwards to enable MASEM. Of these studies, 45 also provided information on the level of socio-economic status (SES) of the students. For the present illustration, I will use these 45 studies. The data and syntax can be found online on http://suzannejak.nl/masem. Based on theory about teacher-student relations (Connell and Wellborn 1991; Pianta 1999), teacher-student relations were considered predictors of engagement and achievement, in which the relation between teacher-student relations and achievement may be mediated by engagement. The hypothetical model representing full mediation of these effects is depicted in Fig. 1.1.

To illustrate the MASEM analysis on these data, I will first fit a fixed effects Stage 1 model. If the Stage 1 model does not fit, the correlation matrices cannot

© The Author(s) 2015
S. Jak, *Meta-Analytic Structural Equation Modelling*,
SpringerBriefs in Research Synthesis and Meta-Analysis,
DOI 10.1007/978-3-319-27174-3_5

be considered homogenous across studies. Study-level heterogeneity can possibly be explained by the average socio-economic status (SES) of the students in the sample. If SES explains the heterogeneity, the fixed effects Stage 1 model should hold within subgroups with high versus low SES. Another approach to account for heterogeneity is to fit a random effects Stage 1 model, allowing for study level variance of the correlation coefficients. The Stage 2 model will be fit on the pooled correlation matrix of the most appropriate Stage 1 analysis.

5.2 Preparing the Data

Before the analysis can start, the data have to be imported in R. The "metaSEM" package in R (Cheung 2014) includes several functions to create a list with correlation matrices for each study. All three functions require the data to be stored in some file. The function `readFullMat()` can be used if the file contains the full correlation matrix for each study. The function `readLowTriMat()` is useful if the file contains the lower triangular (the diagonal and all values below the diagonal) of the correlation matrix for each study. The function `readStackVec()` can be used if the file contains one row with the unique elements of the correlation matrix of each study, and fills the correlation matrices in R by column. The data for the present analyses are stored in the file "Roorda_SES.dat", which is saved in the working directory of R. The first few rows of data look like this:

```
4       1310    -.54    NA      .18     NA      -.29    NA      70
10      427     NA      .64     .29     NA      NA      .23     78
12      123     NA      .29     NA      NA      NA      NA      83
13      66      NA      .29     NA      NA      NA      NA      30
33      179     NA      .22     .08     -.45    -.24    NA      27
```

Here, the first column has an identification number for each study, the second column the sample size, row 3 to 8 have the correlation coefficients and the last row shows the percentage of students with low SES in the sample. NA's represent missing correlation coefficients. Because the `readStackVec()` function requires the diagonal elements of the matrix to be in the datafile as well, this function is not readily useful for this dataset. Therefore, the data is read in with the `read.table()` function. With `head(data)` R will show the beginning of the data, which can be used to inspect whether the data was read in correctly.

```
require("metaSEM")

data <- read.table(file = "Roorda_SES.dat", header = TRUE)

head(data)
```

The next step is to create a list of correlation matrices. First, the number of observed variables is stored in the object `nvar`, and a list with the variable names is created.

```
nvar <- 4
varnames <- c("pos","neg","enga","achiev")
labels <- list(varnames,varnames)
```

The correlation matrices will be stored in the object `cormatrices`. First, an empty list is created that will be filled with one correlation matrix for each study. Because each row has the information of one study, one correlation matrix is created for each row. The coefficients are put in a symmetric matrix using the `vec2symMat()` function. The argument `as.matrix(data[i,3:8])` creates a vector of the elements of row i, column 3 to 8 of the data. The argument `diag = FALSE` indicates that the diagonal elements (1's) are not given in the data, these will be created by the function. The `dimnames()` function gives names to the rows and columns of each correlation matrix.

```
cormatrices <- list()

for (i in 1:nrow(data)){
    cormatrices[[i]] <- vec2symMat(as.matrix(data[i,3:8]),
    diag = FALSE)
    dimnames(cormatrices[[i]]) <- labels
}
```

The previous code creates a four by four correlation matrix for each study. Most studies did not include all variables, and have NA's in the matrix for correlations associated with one or more of the variables. For the TSSEM analysis, we have to put a NA on the corresponding diagonal element of the input matrix if a variable is missing. The following code states that for each correlation matrix, for each row, if the sum of the elements that are NA in that row equals the number of variables minus 1, the diagonal element should be NA.

```
# put NA on diagonal if variable is missing

for (i in 1:length(cormatrices)){
    for (j in 1:nrow(cormatrices[[i]])){
        if (sum(is.na(cormatrices[[i]][j,]))==nvar-1)
            {cormatrices[[i]][j,j] <- NA}
}}
```

Some studies included a variable, but did not report all correlations of the variable with the rest of the variables. For example, the 13th study reported the correlations of Positive interactions with Engagement and Positive interactions with Achievement, but not the correlation between Engagement and Achievement:

```
> cordat[[13]]
          pos neg enga achiev
pos      1.00  NA 0.35   0.01
neg        NA  NA   NA     NA
enga     0.35  NA 1.00     NA
achiev   0.01  NA   NA   1.00
```

For each missing correlation, we have to treat one variable as missing. In this example we would throw away more information if we deleted the variable Positive interactions, than if we deleted Engagement or Achievement. In the following code, for each study, for each missing correlation, the variable that has the least remaining correlations with other variables gets NA on the diagonal.

```
# put NA on diag for var with least present correlations

for (i in 1:length(cordat)){
for (j in 1:nrow(cordat[[i]])){
      for (k in 1:nvar){
          if (is.na(cordat[[i]][j,k]==TRUE)
            &is.na(cordat[[i]][j,j])!=TRUE
            &is.na(cordat[[i]][k,k])!=TRUE){

if(sum(is.na(cordat[[i]])[j,])>sum(is.na(cordat[[i]])[k,]))
    {cordat[[i]][k,k] <- NA}
    if(sum(is.na(cordat[[i]])[j,])<=sum(is.na(cordat[[i]])[k,]))
    {cordat[[i]][j,j] <- NA}
}}}}
```

5.3 Fixed Effects Analysis

The tssem1() function from the metaSEM package can be used to fit the Stage 1 model. As its arguments it uses the list of correlation matrices (cordat), and a vector of sample sizes of the studies (data$N). The argument method = "FEM" indicates that we want to fit the fixed effects model. The results are saved in the object stage1fixed.

```
stage1fixed <- tssem1(my.df=cordat, n=data$N, method="FEM")

summary(stage1fixed)
```

Asking for a summary of the output gives the following results.

```
Coefficients:
        Estimate   Std.Error  z value  Pr(>|z|)
S[1,2] -0.3490484  0.0137214 -25.438 < 2.2e-16 ***
S[1,3]  0.3198412  0.0098717  32.400 < 2.2e-16 ***
S[1,4]  0.1102735  0.0095590  11.536 < 2.2e-16 ***
S[2,3] -0.3061322  0.0163563 -18.716 < 2.2e-16 ***
S[2,4] -0.1441573  0.0076740 -18.785 < 2.2e-16 ***
S[3,4]  0.2399404  0.0137625  17.434 < 2.2e-16 ***
---
Signif. codes:  0 `***' 0.001 `**' 0.01 `*' 0.05 `.' 0.1 ` '
1

Goodness-of-fit indices:
                                     Value
Sample size                      29438.0000
Chi-square of target model         762.2640
DF of target model                  95.0000
p value of target model              0.0000
Chi-square of independence model  3007.5051
DF of independence model           101.0000
RMSEA                                0.1036
SRMR                                 0.1462
TLI                                  0.7559
CFI                                  0.7704
AIC                                572.2640
BIC                               -215.2899
OpenMx status1: 0 ("0" or "1": The optimization is considered
fine.
Other values indicate problems.)
```

The χ^2 of the model with equality constraints on all correlation coefficients across studies is significant $\chi^2_{(95)} = 762.26, p < 0.05$, and the RMSEA is larger than 0.10, indicating bad fit. Based on these fit indices, homogeneity of correlation coefficients has to be rejected. Therefore, I will not continue to fit the Stage 2 model on the pooled correlation matrix of the fixed effects approach.

SES may explain some of the heterogeneity of the correlation coefficients. Therefore, a next step is to fit the fixed effects Stage 1 model separately to studies with less than 50 % students with low SES, and to studies with more than 50 % students with low SES separately. The tssem1() function has the argument cluster to specify the subgroups of studies. In this example there are 21 studies with majority of the sample with high SES, and 24 studies with the majority of the sample with low SES.

```
# Stage 1 FIXED per subgroup

stage1fixed_SES <- tssem1(my.df=cordat, n=data$N,
                 method="FEM", cluster=data$SES>50)

summary(stage1fixed_SES)
```

This gives the following output. The first part of the output (beginning with $ `FALSE'`) is about studies for which the percentage of respondents with low SES was not higher than 50. The second part is about studies with more than 50 % of the respondents with low SES.

```
summary(stage1fixed_SES)
$`FALSE`

Call:
tssem1FEM(my.df = data.cluster[[i]], n = n.cluster[[i]],
cor.analysis = cor.analysis,
    model.name = model.name, suppressWarnings = suppressWarn-
ings)

Coefficients:
         Estimate   Std.Error   z value   Pr(>|z|)
S[1,2] -0.2236276  0.0255706   -8.7455 < 2.2e-16 ***
S[1,3]  0.2326743  0.0296517    7.8469 4.219e-15 ***
S[1,4]  0.1301579  0.0154897    8.4029 < 2.2e-16 ***
S[2,3] -0.1990668  0.0247272   -8.0505 8.882e-16 ***
S[2,4] -0.1321416  0.0086386  -15.2967 < 2.2e-16 ***
S[3,4]  0.3198221  0.0210673   15.1810 < 2.2e-16 ***
---
Signif. codes:  0 '***' 0.001 '**' 0.01 '*' 0.05 '.' 0.1 ' '
1

Goodness-of-fit indices:
                                       Value
Sample size                        17555.0000
Chi-square of target model           165.5335
DF of target model                    41.0000
p value of target model                0.0000
Chi-square of independence model     809.5435
DF of independence model              47.0000
RMSEA                                  0.0603
SRMR                                   0.1325
TLI                                    0.8128
CFI                                    0.8367
AIC                                   83.5335
BIC                                 -235.1634
OpenMx status1: 0 ("0" or "1": The optimization is considered
fine.
 Other values indicate problems.)

$`TRUE`

Call:
tssem1FEM(my.df = data.cluster[[i]], n = n.cluster[[i]],
cor.analysis = cor.analysis,
    model.name = model.name, suppressWarnings = suppressWarn-
ings)

Coefficients:
```

```
          Estimate Std.Error   z value  Pr(>|z|)
S[1,2] -0.411828  0.015644 -26.3247 < 2.2e-16 ***
S[1,3]  0.336493  0.010427  32.2709 < 2.2e-16 ***
S[1,4]  0.099982  0.012384   8.0737 6.661e-16 ***
S[2,3] -0.394882  0.021003 -18.8009 < 2.2e-16 ***
S[2,4] -0.188545  0.016525 -11.4094 < 2.2e-16 ***
S[3,4]  0.194880  0.017918  10.8760 < 2.2e-16 ***
---
Signif. codes:  0 '***' 0.001 '**' 0.01 '*' 0.05 '.' 0.1 ' '
1


Goodness-of-fit indices:
                                     Value
Sample size                     11883.0000
Chi-square of target model        485.8356
DF of target model                 48.0000
p value of target model             0.0000
Chi-square of independence model 2197.9616
DF of independence model           54.0000
RMSEA                               0.1357
SRMR                                0.1367
TLI                                 0.7703
CFI                                 0.7958
AIC                               389.8356
BIC                                35.4582
OpenMx status1: 0 ("0" or "1": The optimization is considered
fine.
  Other values indicate problems.)
```

In the group of studies with high SES (group 'FALSE'), the χ^2 was significant ($\chi^2_{(41)} = 165.53, p < 0.05$), indicating that within studies with high SES, the correlation coefficients are not exactly equal. The RMSEA however is 0.06, which is below the often used 0.08 threshold of satisfactory approximate fit. So, based on the RMSEA it could be concluded that the correlation coefficients are approximately equal within the group of studies with high SES.

In the group of studies with low SES, homogeneity of correlation matrices has to be rejected both based on the significant $\chi^2 (\chi^2_{(48)} = 485.84, p < 0.05)$ and an RMSEA of 0.136. So, as not all heterogeneity could be explained by SES, the random effects approach seems more appropriate.

5.4 Random Effects Analysis

Stage 1: The random effects Stage 1 model can be fit using the method = "REM" argument in the tssem1() function. This should be accompanied by the argument RE.type=, which specifies whether study level variance and covariance should be estimated for all correlation coefficients (the full T^2 matrix from Chap. 3), indicated by RE.type = "Symm". Very often, the amount of observed information is too small to obtain stable estimates of all random effects (Becker 2009; Cheung 2015), leading to an error message when running the model. If this is the case, one may only estimate the study level variance by stating RE.type = "Diag". It is also an option not to estimate study level variance by stating RE.type = "Zero". This would lead to conducting a fixed effects multivariate analysis. In the current example, indeed it was not possible to estimate the full random effects covariance matrix, so RE.type = "Diag" is used.

```
# Stage 1 random

stage1random <- tssem1(my.df=cordat, n=data$N, method="REM",
                       RE.type="Diag")

summary(stage1random)
```

Asking for the summary leads to the following output (to save space, I removed two columns showing the z-values and p-values associated with the parameter estimates).

```
95% confidence intervals: z statistic approximation
Coefficients:
                Estimate    Std.Error       lbound       ubound
value
Intercept1  -0.24271750   0.04095373  -0.32298535  -0.16244966
Intercept2   0.31930285   0.03782976   0.24515788   0.39344782
Intercept3   0.14296055   0.02053875   0.10270534   0.18321577
Intercept4  -0.30862115   0.04449891  -0.39583741  -0.22140488
Intercept5  -0.18035383   0.02565751  -0.23064164  -0.13006603
Intercept6   0.27904891   0.03862889   0.20333767   0.35476015
Tau2_1_1     0.01916654   0.00861737   0.00227682   0.03605627
Tau2_2_2     0.02535841   0.00886709   0.00797924   0.04273758
Tau2_3_3     0.00668849   0.00281654   0.00116817   0.01220882
Tau2_4_4     0.01239430   0.00746887  -0.00224441   0.02703302
Tau2_5_5     0.00704567   0.00370676  -0.00021945   0.01431080
Tau2_6_6     0.01593291   0.00724441   0.00173413   0.03013169

Q statistic on homogeneity of effect sizes: 918.6327
Degrees of freedom of the Q statistic: 95
P value of the Q statistic: 0
Heterogeneity indices (based on the estimated Tau2):
                                     Estimate
Intercept1: I2 (Q statistic)          0.9199
Intercept2: I2 (Q statistic)          0.9426
Intercept3: I2 (Q statistic)          0.7925
Intercept4: I2 (Q statistic)          0.8774
Intercept5: I2 (Q statistic)          0.8016
Intercept6: I2 (Q statistic)          0.9024

Number of studies (or clusters): 45
Number of observed statistics: 101
Number of estimated parameters: 12
Degrees of freedom: 89
-2 log likelihood: -124.8878
OpenMx status1: 0 ("0" or "1": The optimization is considered
fine. Other values indicate problems.)
```

Variable	1	2	3	4
1. Positive relations	1			
2. Negative relations	−0.24	1		
3. Engagement	0.32	−0.31	1	
4. Achievement	0.14	−0.18	0.28	1

Table 5.1 Pooled correlation matrix based on the random effects model

The summary of the result of the random effects analysis shows the Q statistic. The Q statistic is significant ($Q_{(95)} = 918.63$, $p < 0.05$), indicating that there indeed is significant heterogeneity in the correlation matrices. The I^2 of the six correlation coefficients varies between 0.79 and 0.94, so for all coefficients a large part of the variance is at the study level. The averaged correlation coefficients are denoted by "Intercept" and the estimated study level variances of the correlation coefficients are denoted by "Tau2" in the output. The averaged correlation matrix based on the random effects model is shown in Table 5.1.

The `tssem1()` function also returns the asymptotic variance covariance matrix for Stage 1 estimates. This matrix will be used as the weight matrix when estimating the Stage 2 model using WLS-estimation. The asymptotic variance covariance matrix for this example can be viewed using `vcov(stage1random)`.

Stage 2: The structural model that we are going to fit to the pooled correlation matrix is the model that was also used as an example in Chap. 1, see Fig. 1.1. The specification of any structural model in the metaSEM package is done using three matrices (the RAM-formulation, McArdle and McDonald 1984). Matrix A specifies all regression coefficients in the model, Matrix S specifies all variances and covariances in the model, and matrix F indicates which variables are observed and which variables are latent. If all variables are observed, which is the case for this path model, the F matrix is not needed. The model matrices always have number of rows and number of columns equal to the number of (observed + latent) variables in the model. The A-matrix of the current example is thus a four by four matrix, in which three regression coefficients are specified, β_{31}, β_{32}, and β_{43}. The function `create.mxMatrix()` facilitates the specification of the matrices that the program OpenMx (by which the model is really fitted) needs. The A-matrix for the current example is created by the following code.

```
A <- create.mxMatrix(
                    c( 0,0,0,0,
                       0,0,0,0,
                       "0.1*b31","0.1*b32",0,0,
                       0,0,"0.1*b43",0),
                    type = "Full",
                    nrow = 4,
                    ncol = 4,
                    byrow = TRUE,
                    name = "A",
                    dimnames = list(varnames,varnames))
```

If a number is specified in the A-matrix, it indicates that the corresponding parameter is not estimated but fixed (fixed at the given number, zero in this case).

If it is not a number, but for example "0.1*b31", the parameter is given a starting value of 0.1 and it gets the label b31. A starting value is the value that the program will use as a starting point for the iterative estimation procedure. Different starting values should lead to the same parameter estimate, and are thus quite arbitrary, although some starting values may lead to problems such as a non-positive definite model implied correlation matrix. See Bollen (1989) or Kline (2011) for some guidelines on starting values. Note that by default, the information about the parameters is read in column wise by the `create.mxMatrix()` function. This can be changed using the `byrow = TRUE` argument.

The created matrices for the regression coefficients are a matrix indicating the labels of the parameters, a matrix with the starting values of the parameters and a matrix indicating whether the parameter is freely estimated (indicated by `TRUE`) or not (indicated by `FALSE`). Each parameter could also be given a lower and upper bound for the estimate, but this is not often needed. This is what the object A entails:

```
FullMatrix 'A'

$labels
          pos    neg    enga   achiev
pos       NA     NA     NA     NA
neg       NA     NA     NA     NA
enga      "b31"  "b32"  NA     NA
achiev    NA     NA     "b43"  NA

$values
          pos neg enga achiev
pos       0.0 0.0  0.0       0
neg       0.0 0.0  0.0       0
enga      0.1 0.1  0.0       0
achiev    0.0 0.0  0.1       0

$free
           pos    neg    enga   achiev
pos        FALSE  FALSE  FALSE  FALSE
neg        FALSE  FALSE  FALSE  FALSE
enga       TRUE   TRUE   FALSE  FALSE
achiev     FALSE  FALSE  TRUE   FALSE

$lbound: No lower bounds assigned.

$ubound: No upper bounds assigned.
```

The S-matrix contains the information about variances (on the diagonal) and covariances (off-diagonal) in the model. In the present model there are two variances of exogenous variables, one covariance between exogenous variables, and two residual variances for endogenous variables. The S-matrix is also a four by four matrix, and because it is symmetrical, we only have to provide the lower triangular of the matrix (columnwise). Here I gave the labels p_{11} to p_{44} for the variances, with startvalues of 1, and the label p_{21} with a startvalue of 0.1 for the covariance between the first two variables.

```
S <- create.mxMatrix(
                c("1*p11",
                  ".1*p21","1*p22",
                  0,0,"1*p33",
                  0,0,0,"1*p44"),
                type="Symm",
                byrow = TRUE,
                name="S",
                dimnames = list(varnames,varnames))
```

The resulting object S looks like this:

```
SymmMatrix 'S'

$labels
          pos   neg   enga   achiev
pos      "p11" "p21" NA     NA
neg      "p21" "p22" NA     NA
enga     NA    NA    "p33"  NA
achiev   NA    NA    NA     "p44"

$values
          pos neg enga achiev
pos      1.0 0.1   0      0
neg      0.1 1.0   0      0
enga     0.0 0.0   1      0
achiev   0.0 0.0   0      1

$free
           pos   neg   enga  achiev
pos       TRUE   TRUE  FALSE  FALSE
neg       TRUE   TRUE  FALSE  FALSE
enga      FALSE FALSE  TRUE   FALSE
achiev    FALSE FALSE  FALSE  TRUE

$lbound: No lower bounds assigned.

$ubound: No upper bounds assigned.
```

Using these two matrices, the hypothesized model can be fit to the data using the `tssem2()` function. This function needs as its arguments the object with the Stage 1 results (either from a fixed or a random effects analysis), and the A- and S-matrices. Two additional arguments are given. The argument `diag. constraints = TRUE` ensures that the diagonal of the model implied correlation matrix always consists of 1's during estimation. This is required because the input matrix is a correlation matrix and not a covariance matrix. Another option exists (used with `diag.constraints = FALSE`), but is only appropriate when there are no mediators in the model, and it has the downside of not providing estimates of the residual variances of endogenous variables.

The argument `intervals = "LB"` is used to ask for likelihood based confidence intervals (Neale and Miller 1997). Using likelihood based confidence intervals for significance testing is not only sometimes better than using standard error based confidence intervals, for example when testing indirect effects (Cheung 2009), but they are also the only option when `diag.constraints = TRUE` is used.

```
stage2 <- tssem2(stage1random, Amatrix=A, Smatrix=S,
                 diag.constraints=TRUE, intervals="LB")
```

Asking for a summary gives the following output (again, I removed the columns for the z- and p-values).

```
95% confidence intervals: Likelihood-based statistic
Coefficients:
     Estimate Std.Error    lbound    ubound
b43   0.34853        NA   0.28826   0.41002
b32  -0.29864        NA  -0.38116  -0.21513
b31   0.27440        NA   0.20119   0.34651
p44   0.87853        NA   0.83197   0.91690
p33   0.79622        NA   0.73118   0.85067
p22   1.00000        NA   1.00000   1.00000
p21  -0.23980        NA  -0.32011  -0.15949
p11   1.00000        NA   1.00000   1.00000

Goodness-of-fit indices:
                                                        Value
Sample size                                        29438.0000
Chi-square of target model                            11.1273
DF of target model                                     2.0000
p value of target model                                0.0038
Number of constraints imposed on "Smatrix"             4.0000
DF manually adjusted                                   0.0000
Chi-square of independence model                     276.7660
DF of independence model                               6.0000
RMSEA                                                  0.0125
SRMR                                                  0.0447
TLI                                                   0.8989
CFI                                                  0.9663
AIC                                                   7.1273
BIC                                                  -9.4528
OpenMx status1: 0 ("0" or "1": The optimization is considered
fine.
Other values indicate problems.)
```

The χ^2 of the hypothesized path model is significant ($\chi^2_{(2)} = 11.13, p < 0.05$, so exact fit is rejected. However, the RMSEA of 0.013 indicated close approximate fit, and the CFI of 0.97 also indicates satisfactory fit of the model. The parameter estimates from the A- and S-matrix are all significantly different from zero, as zero is not included in the 95 % confidence intervals. Figure 5.1 shows the path model with

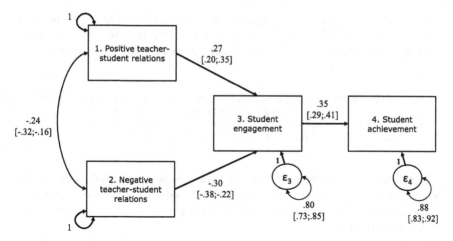

Fig. 5.1 Path model with parameter estimates and 95 % confidence intervals

the parameter estimates and 95 % confidence intervals. This figure is created manually, but the metaSEM package also includes a function to create a graphical display of a model. The following code generates a graph with the parameter estimates.

```
# plot

install.packages("semPlot") # install package
require("semPlot")          # load the package

my.plot <- meta2semPlot(stage2)
semPaths(my.plot, whatLabels="est", layout = "tree2")
```

Besides the direct effects, the indirect effects may be of interest. The indirect effect is equal to the product of the direct effects that constitute the indirect effect. For example, the indirect effect of positive relations to achievement in this example is equal to $0.27 * 0.35 = 0.10$. The significance of indirect effects can also be tested using likelihood based confidence intervals. The following code can be used to fit the Stage 2 model and estimate the likelihood based confidence intervals for the indirect effects.

```
# Stage 2 model with indirect effects

stage2 <- tssem2(stage1random, Amatrix=A, Smatrix=S,
                diag.constraints=TRUE, intervals="LB",
                mx.algebras=list(
                    Indpos=mxAlgebra(b31*b43,name="Indpos"),
                    Indneg=mxAlgebra(b32*b43,name="Indneg"))
                )
```

The summary of the stage2 object then provides the estimates of the indirect effects with the 95 % likelihood based confidence intervals. As zero is not included in both intervals, both indirect effects can be considered significant.

```
mxAlgebras objects (and their 95% likelihood-based CIs):
                lbound      Estimate        ubound
Indpos[1,1]   0.06851276   0.09563552   0.12429360
Indneg[1,1]  -0.13849366  -0.10408579  -0.07210687
```

Significant indirect effects indicate that there is at least partial mediation. If there are no significant direct effects, there is full mediation. The two direct effects from positive and negative relations to achievement are estimated by specifying the A-matrix as follows:

```
# model with indirect and direct effects

A2 <- create.mxMatrix(
                 c( 0,0,0,0,
                    0,0,0,0,
                    "0.1*b31","0.1*b32",0,0,
                    "0.1*b41","0.1*b42","0.1*b43",0),
                 type = "Full",
                 nrow = 4,
                 ncol = 4,
                 byrow = TRUE,
                 name = "A",
                 dimnames = list(varnames,varnames))

stage2_2 <- tssem2(stage1random, Amatrix=A2, Smatrix=S,
                 diag.constraints=TRUE,intervals="LB",
                 mx.algebras=list(
                    Indpos=mxAlgebra(b31*b43,name="Indpos"),
                    Indneg=mxAlgebra(b32*b43,name="Indneg"))
                 )
```

The resulting model is a saturated model, which means that degrees of freedom are zero, and the fit of the model cannot be evaluated. However, we can still evaluate the significance of the parameters. The direct effect of positive relations on achievement is estimated as $\beta = 0.044$, with 95 % confidence interval running from -0.013 to 0.098. The direct effect is not significant, so as expected, the effect of positive relations on achievement is said to be fully mediated by engagement. For negative relations, the direct effect is estimated as -0.097 with 95 % CI running between -0.160 and -0.031, so the effect of negative relations is said to be partially mediated by engagement.

5.5 Random Effects Subgroup Analysis

It may be of substantive interest to compare the parameter estimates of the Stage 2 model across subgroups of studies. For example, one might want to investigate whether and how the regression coefficients of the structural model differ across studies with a majority of the sample defined as having low SES, and studies with a majority of the sample defined as having high SES. As we already know from the fixed effects analysis that not all heterogeneity is explained by SES, a random effects model in each subgroup seems suitable. A potential problem with subgroup analysis is that the number of studies within each subgroup may become quite small. Indeed, in this example there are respectively 24 and 21 studies included in the two groups, so the results should be interpreted with caution.

In the Stage 1 function `tssem1()`, the `cluster` option is not available for random effects analysis. Therefore, we will create two separate lists of correlation matrices and sample sizes for the two subgroups of studies. Using the following code, we create these lists, by selecting the studies with a value on SES that is higher than 50 (a higher value indicates more children with low SES). Then, we run the Stage 1 model separately in the two groups of studies.

```
# majority low SES
cordat_lo <- cordat[data$SES>50]
N_lo <- data$N[data$SES>50]

# majority high SES
cordat_hi <- cordat[data$SES<=50]
N_hi <- data$N[data$SES<=50]


stage1random_lo <- tssem1(my.df=cordat_lo, n=N_lo,
                          method="REM", RE.type="Diag")

stage1random_hi <- tssem1(my.df=cordat_hi, n=N_hi,
                          method="REM", RE.type="Diag")
```

Then, using the A- and S-matrices we already created, we can fit the structural model to the averaged correlation matrices in the two subgroups separately.

```
stage2_lo <- tssem2(stage1random_lo, Amatrix=A, Smatrix=S,
                    diag.constraints=TRUE, intervals="LB")

stage2_hi <- tssem2(stage1random_hi, Amatrix=A, Smatrix=S,
                    diag.constraints=TRUE, intervals="LB")
summary(stage2_lo)
summary(stage2_hi)
```

The (truncated) output of the summary is given below.

```
summary(stage2_lo)

95% confidence intervals: Likelihood-based statistic
Coefficients:
     Estimate Std.Error    lbound    ubound
b43   0.28762        NA   0.21216   0.36442
b32  -0.30694        NA  -0.41951  -0.19242
b31   0.26516        NA   0.16390   0.36298
p44   0.91728        NA   0.86728   0.95506
p33   0.78190        NA   0.69835   0.84998
p22   1.00000        NA   1.00000   1.00000
p21  -0.32918        NA  -0.42137  -0.23700
p11   1.00000        NA   1.00000   1.00000

Goodness-of-fit indices:
                                                      Value
Sample size                                      11883.0000
Chi-square of target model                           6.2731
DF of target model                                   2.0000
p value of target model                              0.0434
Number of constraints imposed on "Smatrix"           4.0000
DF manually adjusted                                 0.0000
Chi-square of independence model                   201.4439
DF of independence model                             6.0000
RMSEA                                                0.0134
SRMR                                                 0.0411
TLI                                                  0.9344
CFI                                                  0.9781
AIC                                                  2.2731
BIC                                                -12.4926
OpenMx status1: 0 ("0" or "1": The optimization is considered
fine.
Other values indicate problems.)

> summary(stage2_hi)
```

```
95% confidence intervals: Likelihood-based statistic
Coefficients:
      Estimate Std.Error    lbound      ubound
b43   0.413576        NA  0.326113    0.502127
b32  -0.248525        NA -0.345284   -0.150065
b31   0.250379        NA  0.157087    0.343685
p44   0.828954        NA  0.747939    0.893711
p33   0.855686        NA  0.781408    0.913294
p22   1.000000        NA  1.000000    1.000000
p21  -0.159580        NA -0.268583   -0.050578
p11   1.000000        NA  1.000000    1.000000

Goodness-of-fit indices:
                                                    Value
Sample size                                   17555.0000
Chi-square of target model                        9.4817
DF of target model                                2.0000
p value of target model                           0.0087
Number of constraints imposed on "Smatrix"        4.0000
DF manually adjusted                              0.0000
Chi-square of independence model                122.9021
DF of independence model                          6.0000
RMSEA                                             0.0146
SRMR                                              0.0549
TLI                                               0.8080
CFI                                               0.9360
AIC                                               5.4817
BIC                                             -10.0645
OpenMx status1: 0 ("0" or "1": The optimization is considered
fine.
Other values indicate problems.)
```

The path model shows acceptable fit according to the RMSEA and CFI in the low-SES studies, $\chi^2_{(2)} = 6.27, p < 0.05$, RMSEA $= 0.013$, CFI $= 0.98$ as well as in the high-SES studies, $\chi^2_{(2)} = 9.48, p < 0.05$, RMSEA $= 0.015$, CFI $= 0.94$. The parameter estimates for both groups are shown in Table 5.2. Some estimates seem to be different across the two groups. For example, the effect of Negative interactions on Engagement (β_{32}) seems to be stronger for students with low SES, and the effect of Engagement on Achievement (β_{43}) seems to be stronger for students with high SES. As the confidence intervals of these estimates in the two groups overlap, we cannot be certain that the effects are significantly different across samples with high and low SES. This could be tested by fitting a multigroup Stage 2 model, and constraining the parameters to be equal across groups. If the χ^2 increases significantly when adding equality constraints across groups, the parameters are significantly different across groups. These analyses cannot be performed using the functions in the metaSEM-package, but need specification of the model in openMx directly. Doing the analyses in OpenMx showed that only the effect of Engagement of Achievement is significantly

Table 5.2 Parameter estimates and 95 % confidence intervals for studies with High and Low SES

Parameter	Estimate [lower bound; upper bound]	
	Low SES	High SES
β_{31}	0.27 [0.16; 0.36]	0.25 [0.16; 0.34]
β_{32}	−0.31 [−0.42; −0.19]	−0.25 [−0.35; −0.15]
β_{43}	0.29 [0.21; 0.36]	0.41 [0.33; 0.50]
ψ_{21}	−0.33 [−0.42; −0.24]	−0.16 [−0.27; −0.05]
ψ_{33}	0.78 [0.70; −0.85]	0.86 [0.78; 0.91]
ψ_{44}	0.92 [0.87; 0.89]	0.83 [0.75; 0.89]

different across groups ($\chi^2_{(1)} = 4.51, p < 0.05$). So, apparently, the effect of Engagement on Achievement is higher for children with high SES. The code that was used to test the difference in effects across groups can be found online at http://suzannejak.nl/masem.

References

Becker, B. J. (2009). Model-based meta-analysis. In H. Cooper, L. V. Hedges, & J. C. Valentine (Eds.), *The handbook of research synthesis and meta-analysis* (2nd ed., pp. 377–398). New York: Russell Sage Foundation.

Bollen, K. A. (1989). *Structural equations with latent variables*. New York: Wiley.

Cheung, M. W.-L. (2009). Comparison of methods for constructing confidence intervals of standardized indirect effects. *Behavior Research Methods, 41*(2), 425–438.

Cheung, M. W.-L. (2014). Fixed- and random-effects meta-analytic structural equation modeling: examples and analyses in R. *Behavior Research Methods, 46*, 29–40.

Cheung, M. W.-L. (2015). metaSEM: An R package for meta-analysis using structural equation modeling. *Frontiers in Psychology 5*, 1521. doi: 10.3389/fpsyg.2014.01521

Connell, J. P., & Wellborn, J. G. (1991). Competence, autonomy, and relatedness: A motivational analysis of self-system processes. In M. Gunnar & A. Sroufe (Eds.), *The Minnesota symposium on child development: Self-processes and development* (Vol. 23, pp. 43–77). Hillsdale: Lawrence Erlbaum Associates.

Kline, R. B. (2011). *Principles and practice of structural equation modeling*. New York: Guilford.

McArdle, J. J., & McDonald, R. P. (1984). Some algebraic properties of the reticular action model for moment structures. *British Journal of Mathematical and Statistical Psychology, 37*(2), 234–251.

Neale, M. C., & Miller, M. B. (1997). The use of likelihood based confidence intervals. *Genetics, 27*, 113–120.

Pianta, R. C. (1999). *Enhancing relationships between children and teachers*. Washington, DC: American Psychological Association.

Roorda, D. L., Koomen, H. M., Spilt, J. L., & Oort, F. J. (2011). The influence of affective teacher–student relationships on students' school engagement and achievement a meta-analytic approach. *Review of Educational Research, 81*(4), 493–529.

Chapter 6
Fitting a Factor Model with the Two-Stage Approach

Abstract In this chapter I will illustrate fitting a factor model within a MASEM analysis using the metaSEM package. The data come from a meta-analysis performed by Fan et al. (Personality and Individual Differences 48(7):781–785, 2010), who collected correlation matrices of the 8 subscales of a test to measure "Emotional intelligence" from 19 studies. The preparation of the data, and the fixed and random effects Stage 1 analyses are explained step by step. Next, the Stage 2 factor model is fit to the pooled correlation matrix from the random effects Stage 1 analysis. All steps that have to be taken to perform the analyses are discussed, as well as the relevant output.

Keywords Meta-analytic structural equation modeling · MetaSEM · Factor model · Emotional intelligence · MSCEIT · Fixed effects · Random effects

6.1 Introduction

Fan et al. (2010) used meta-analytic factor analysis to investigate the factor structure of a measurement instrument of emotional intelligence, the Mayer-Salovey-Caruso Emotional Intelligence Test Version 2.0 (MSCEIT). Emotional intelligence is defined as a set of skills hypothesized to contribute to the accurate appraisal and expression of emotion, the effective regulation of emotion, and the use of feelings to motivate, plan, and achieve in one's life (Salovey and Mayer 1989). The MSCEIT consists of 8 subscales. Previous research on the factor structure of the MSCEIT lead to contradictory results, and a MASEM made it possible to compare the fit of several proposed factor models on the aggregated data across 19 studies. Based on these analyses, a three-factor model was found to have the best fit. In this section I will replicate the fixed effects analysis of Fan et al. and additionally run a random effects MASEM. The data and script to replicate the analyses can be found on my website (http://suzannejak.nl/masem).

© The Author(s) 2015
S. Jak, *Meta-Analytic Structural Equation Modelling*,
SpringerBriefs in Research Synthesis and Meta-Analysis,
DOI 10.1007/978-3-319-27174-3_6

6.2 Preparing the Data

Fan et al. collected 19 correlation matrices from different studies. Most of the studies reported all correlations between the 8 scales of the MSCEIT, for some studies the correlation had to be deduced from other information (see Fan et al.) and for two studies one and two variables were missing. The correlation matrices are collected in a text file, "fan_msceit.dat", which contains the lower triangular of the matrix in each study. This is a part of the file:

```
1
.34    1
.36    .33    1
.24    .22    .17    1
.23    .12    .32    .31    1
.14    .11    .26    .19    .43    1
.14    -.06   .22    .18    .23    .34    1
.11    -.11   .16    .31    .43    .48    .47    1

1
.406   1
.312   .376   1
.373   .450   .375   1
.258   .297   .189   .227   1
0      0      0      0      0      NA
.309   .372   .270   .324   .361   0      1
.322   .388   .282   .337   .377   0      .511   1
```

The function `readLowTriMat()` can be used to store these matrices in a list that can serve as input for the analysis. The function takes the filename and the number of variables per study as arguments, and then creates a list of correlation matrices. If variables are missing in some studies, this should be indicated by NA on the diagonal. The second matrix shown above does not contain information about the sixth variable, the NA on the diagonal ensures that the associated rows and columns will be filtered out during the analysis (so it does not matter what values are given for the missing correlations). The next two lines of code create the list of matrices and a vector with the associated sample sizes. The argument skip = 1 is needed because the first line of the file contains copyright information, and should be skipped by the function.

```
cordat <- readLowTriMat(file = "fan_msceit.dat", no.var = 8,
                        skip = 1)

N <- c(5000,457,412,655,150,450,138,237,314,405,
       375,239,260,266,209,84,192,523,198)
```

6.3 Fixed Effects Analysis

The `tssem()` function is used to estimate the pooled correlation matrix under the fixed effects model.

```
stage1fixed <- tssem1(my.df = cordat, n = N,
                      method = "FEM")
summary(stage1fixed)
```

Leading to this output:

```
Coefficients.
        Estimate Std.Error z value  Pr(>|z|)
S[1,2] 0.3690033 0.0084335   43.755 < 2.2e-16 ***
S[1,3] 0.3164490 0.0088628   35.706 < 2.2e-16 ***
S[1,4] 0.3291198 0.0087740   37.511 < 2.2e-16 ***
S[1,5] 0.1857085 0.0094337   19.686 < 2.2e-16 ***
S[1,6] 0.1944074 0.0094758   20.516 < 2.2e-16 ***
S[1,7] 0.2135400 0.0093234   22.904 < 2.2e-16 ***
S[1,8] 0.2235946 0.0092685   24.124 < 2.2e-16 ***
S[2,3] 0.3643780 0.0085434   42.650 < 2.2e-16 ***
S[2,4] 0.3263959 0.0087876   37.143 < 2.2e-16 ***
S[2,5] 0.2337159 0.0092278   25.327 < 2.2e-16 ***
S[2,6] 0.2018553 0.0094436   21.375 < 2.2e-16 ***
S[2,7] 0.2428587 0.0092039   26.387 < 2.2e-16 ***
S[2,8] 0.2215793 0.0092865   23.860 < 2.2e-16 ***
S[3,4] 0.3655632 0.0085451   42.781 < 2.2e-16 ***
S[3,5] 0.3333873 0.0087821   37.962 < 2.2e-16 ***
S[3,6] 0.2779142 0.0091839   30.261 < 2.2e-16 ***
S[3,7] 0.3418394 0.0087278   39.167 < 2.2e-16 ***
S[3,8] 0.3173547 0.0088668   35.791 < 2.2e-16 ***
S[4,5] 0.2492657 0.0092489   26.951 < 2.2e-16 ***
S[4,6] 0.2572274 0.0093092   27.632 < 2.2e-16 ***
S[4,7] 0.3242843 0.0088399   36.684 < 2.2e-16 ***
S[4,8] 0.3189749 0.0088592   36.005 < 2.2e-16 ***
S[5,6] 0.4907931 0.0075245   65.226 < 2.2e-16 ***
S[5,7] 0.3641924 0.0085319   42.686 < 2.2e-16 ***
S[5,8] 0.3272308 0.0087472   37.410 < 2.2e-16 ***
S[6,7] 0.3211214 0.0088663   36.218 < 2.2e-16 ***
S[6,8] 0.3489128 0.0086744   40.223 < 2.2e-16 ***
S[7,8] 0.5065221 0.0072959   69.425 < 2.2e-16 ***
---
Signif. codes:  0 '***' 0.001 '**' 0.01 '*' 0.05 '.' 0.1 ' '
1
```

```
Goodness-of-fit indices:
                                         Value
Sample size                         10564.0000
Chi-square of target model           1818.8709
DF of target model                    484.0000
p value of target model                 0.0000
Chi-square of independence model    19130.4290
DF of independence model              512.0000
RMSEA                                   0.0704
SRMR                                    0.1267
TLI                                     0.9242
CFI                                     0.9283
AIC                                   850.8709
BIC                                 -2665.4894
OpenMx status1: 0 ("0" or "1": The optimization is considered
fine.
 Other values indicate problems.)
```

The degrees of freedom are equal to the number of observed correlation coefficients minus the number of estimated correlation coefficients. There are 17 observed complete correlation matrices with 8 * 7/2 = 28 correlation coefficients each. One study missed one variable, and has 7 * 6/2 = 21 coefficients, and one study missed 2 variables and has 6 * 5/2 = 15 observed coefficients. So, in total there are 17 * 28 + 21 + 15 = 512 observed correlation coefficients. The model has 28 parameters, which are the correlation coefficients that are assumed to be equal across studies. Hence, degrees of freedom are 512 − 28 = 484. This calculation leads to the correct number of degrees of freedom, but in reality the diagonal elements of the observed correlation matrices are also counted as observed statistics, and a diagonal matrix is also estimated for each observed matrix (see Eq. 2.7 in Chap. 2). Because the number of observed diagonal elements is equal to the number of estimated diagonal elements, degrees of freedom do not change by evaluating the diagonal elements.

The chi-square is significant ($\chi^2_{(484)} = 1818.87$, $p < 0.05$), exact fit of the Stage 1 model does not hold, indicating that exact homogeneity of the correlation coefficients across studies is rejected. The RMSEA of 0.07 however shows acceptable approximate fit, which could serve as an indication that homogeneity holds

Table 6.1 Pooled correlation matrix of the research variables from the fixed effects analysis

	1	2	3	4	5	6	7	8
1. Faces	1							
2. Pictures	0.37	1						
3. Facilitation	0.32	0.36	1					
4. Sensations	0.33	0.33	0.37	1				
5. Changes	0.19	0.23	0.33	0.25	1			
6. Blends	0.19	0.20	0.28	0.26	0.49	1		
7. Emotional management	0.21	0.24	0.34	0.32	0.36	0.32	1	
8. Emotional relations	0.22	0.22	0.32	0.32	0.33	0.35	0.51	1

approximately, and the pooled correlation matrix from the fixed effects analysis could be used to fit the structural model. Table 6.1 shows the rounded parameter estimates in matrix form. These coefficients can be extracted from the output with `coef(stage1fixed)`.

6.4 Random Effects Analysis

Stage 1: A random effects analysis also seems appropriate for these data. If the heterogeneity of the correlation coefficients is not substantial, the results will not be very different from the fixed effects analysis. The following code will run the random effects Stage 1 analysis. As it was not possible to estimate the study-level covariance, the random effects type "`Diag`" is used.

```
stage1random <- tssem1(my.df = cordat, n = N, method = "REM",
                       RE.type = "Diag")

summary(stage1random)
```

To save space, the raw output is not shown here. The Q-statistic is significant ($Q_{(484)} = 2061.08$), so homogeneity is rejected based on this test. The I^2 of the correlation coefficients range between 0.19 and 0.88 indicating substantial heterogeneity. Table 6.2 shows the pooled correlation matrix from the random effects analysis (with the I^2 values above the diagonal).

The correlation coefficients are somewhat different from the fixed effects estimates. Another difference is in the asymptotic variance covariance matrix of these correlation coefficients that will be used as a weight matrix in the Stage 2 analysis. The asymptotic variance from the random effects analysis will be larger, leading to larger confidence intervals around the Stage 2 estimates.

Stage 2: I am going to fit the structural model to the pooled random effects matrix from Stage 1. Figure 6.1 shows the 3-factor structure that will be fitted to these data. The specification of the parameter matrices for the Stage 2 model does not differ between the random or fixed approach. In the illustration of the path model in Chap. 5, I already introduced the A-matrix with regression coefficients and the S-matrix with variances and covariances. These matrices feature in the factor model as well. The A-matrix contains the factor loadings (λ's in Fig. 6.1), and matrix S contains the residual variances (θ's in Fig. 6.1) as well as the factor variances and covariances (φ's in Fig. 6.1). For factor analysis, a third matrix is needed, which is a matrix that indicates which variables are observed and which variables are latent. This is matrix F. In the current example, we have 8 observed variables and 3 factors. Therefore both the A-matrix and the S-matrix will have 11 rows and 11 columns. The F-matrix will have 8 rows and 11 columns. Matrix F is a selection matrix that filters out the latent variables, it is an identity matrix with the rows associated with the latent variables removed. In the current

Table 6.2 Pooled correlations (below diagonal) and I^2 (above the diagonal) of the research variables from the random effects analysis

	1	2	3	4	5	6	7	8
1. Faces	1	0.82	0.68	0.19	0.47	0.41	0.41	0.42
2. Pictures	0.37	1	0.59	0.37	0.30	0.46	0.57	0.46
3. Facilitation	0.31	0.32	1	0.65	0.78	0.78	0.84	0.68
4. Sensations	0.32	0.31	0.33	1	0.77	0.86	0.74	0.77
5. Changes	0.22	0.21	0.27	0.27	1	0.88	0.86	0.84
6. Blends	0.20	0.20	0.24	0.25	0.45	1	0.76	0.74
7. Emotional management	0.21	0.21	0.30	0.28	0.28	0.28	1	0.87
8. Emotional relations	0.22	0.19	0.27	0.31	0.31	0.32	0.45	1

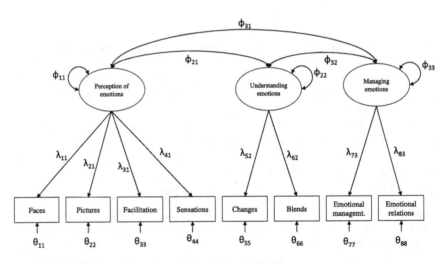

Fig. 6.1 Three factor model on the subscales of the MSCEIT

example, we put the observed variables first, the F matrix can be created using the `create.Fmatrix()` function directly:

```
F <- create.Fmatrix(c(1,1,1,1,1,1,1,1,0,0,0), name="F")
```

Next, we need the A-matrix. I am going to create the A-matrix in steps. First I will create a 8 by 3 matrix lambda, which has the factor loadings.

```
lambda <- matrix(
            c("0.3*L11",0,0,
              "0.3*L21",0,0,
              "0.3*L31",0,0,
              "0.3*L41",0,0,
              0,"0.3*L52",0,
              0,"0.3*L62",0,
              0,0,"0.3*L73",
              0,0,"0.3*L83"),
            nrow=8,
            ncol=3,
            byrow = TRUE)
```

Like the matrices in the path model, if a number is specified in the lambda matrix, it indicates that the factor loading is not estimated but fixed (fixed at the given number, zero in this case). If it is not a number, but for example "0.3 * L11", the parameter is given a starting value of 0.3 and it gets the label "L11". To correctly fix and free elements it may help to think of the columns of lambda as being associated with the common factors and the rows as being associated with the indicators. For example, if indicator number three loads on the first factor (or, the third indicator variable regresses on the first factor), we specify a free parameter for the element in the third row, first column ("0.3 * L31"). Note that the `matrix()` function fills in the values column wise by default, so we use the argument `byrow = TRUE`. The object lambda looks like shown below.

```
> lambda
        [,1]       [,2]         [,3]
[1,] "0.3*L11" "0"        "0"
[2,] "0.3*L21" "0"        "0"
[3,] "0.3*L31" "0"        "0"
[4,] "0.3*L41" "0"        "0"
[5,] "0"       "0.3*L52"  "0"
[6,] "0"       "0.3*L62"  "0"
[7,] "0"       "0"        "0.3*L73"
[8,] "0"       "0"        "0.3*L83"
```

The A-matrix should be an 11 by 11 matrix, in which the factor loadings are in rows 1–8 (associated with the observed variables) and columns 9–11 (associated with the factors). The rest of the matrix should consist of zeros, as there are no other regression coefficients than factor loadings in the model. The zeros can be added to the A matrix by adding a 8 by 8 matrix to the left of lambda and consequently a 3 by 11 matrix with zeros below using the `cbind()` and `rbind()` functions. Next, the `as.mxMatrix()` function is used to create the matrices that are used by OpenMx, which are a matrix indicating the labels of the parameters, a matrix with the starting values of the parameters and a matrix indicating whether the parameter is freely estimated (indicated by `TRUE`) or not (indicated by `FALSE`).

```
A <- rbind(cbind(matrix(0,ncol=8,nrow=8),lambda),
           matrix(0, nrow=3, ncol=11))

A <- as.mxMatrix(A)

# not required but it helps to provide labels
dimnames(A) <- list(
c("face","pict","faci","sens","chen","blen","emma","emre","F1
","F2","F3"),
c("face","pict","faci","sens","chen","blen","emma","emre","F1
","F2","F3"))
```

The resulting A-matrices look as follows.

```
FullMatrix 'A'

$labels
     face pict faci sens chen blen emma emre F1     F2     F3
face NA   NA   NA   NA   NA   NA   NA   NA   "L11" NA     NA
pict NA   NA   NA   NA   NA   NA   NA   NA   "L21" NA     NA
faci NA   NA   NA   NA   NA   NA   NA   NA   "L31" NA     NA
sens NA   NA   NA   NA   NA   NA   NA   NA   "L41" NA     NA
chen NA   NA   NA   NA   NA   NA   NA   NA   NA    "L52" NA
blen NA   NA   NA   NA   NA   NA   NA   NA   NA    "L62" NA
emma NA   NA   NA   NA   NA   NA   NA   NA   NA    NA    "L73"
emre NA   NA   NA   NA   NA   NA   NA   NA   NA    NA    "L83"
F1   NA   NA   NA   NA   NA   NA   NA   NA   NA    NA     NA
F2   NA   NA   NA   NA   NA   NA   NA   NA   NA    NA     NA
F3   NA   NA   NA   NA   NA   NA   NA   NA   NA    NA     NA

$values
     face pict faci sens chen blen emma emre F1 F2  F3
face 0    0    0    0    0    0    0    0 0.3 0.0 0.0
pict 0    0    0    0    0    0    0    0 0.3 0.0 0.0
faci 0    0    0    0    0    0    0    0 0.3 0.0 0.0
sens 0    0    0    0    0    0    0    0 0.3 0.0 0.0
chen 0    0    0    0    0    0    0    0 0.0 0.3 0.0
blen 0    0    0    0    0    0    0    0 0.0 0.3 0.0
emma 0    0    0    0    0    0    0    0 0.0 0.0 0.3
emre 0    0    0    0    0    0    0    0 0.0 0.0 0.3
F1   0    0    0    0    0    0    0    0 0.0 0.0 0.0
F2   0    0    0    0    0    0    0    0 0.0 0.0 0.0
F3   0    0    0    0    0    0    0    0 0.0 0.0 0.0

$free
      face  pict  faci  sens  chen  blen  emma  emre    F1    F2    F3
face FALSE FALSE FALSE FALSE FALSE FALSE FALSE FALSE  TRUE FALSE FALSE
pict FALSE FALSE FALSE FALSE FALSE FALSE FALSE FALSE  TRUE FALSE FALSE
faci FALSE FALSE FALSE FALSE FALSE FALSE FALSE FALSE  TRUE FALSE FALSE
sens FALSE FALSE FALSE FALSE FALSE FALSE FALSE FALSE  TRUE FALSE FALSE
chen FALSE FALSE FALSE FALSE FALSE FALSE FALSE FALSE FALSE  TRUE FALSE
blen FALSE FALSE FALSE FALSE FALSE FALSE FALSE FALSE FALSE  TRUE FALSE
emma FALSE FALSE FALSE FALSE FALSE FALSE FALSE FALSE FALSE FALSE  TRUE
emre FALSE FALSE FALSE FALSE FALSE FALSE FALSE FALSE FALSE FALSE  TRUE
F1   FALSE FALSE FALSE FALSE FALSE FALSE FALSE FALSE FALSE FALSE FALSE
F2   FALSE FALSE FALSE FALSE FALSE FALSE FALSE FALSE FALSE FALSE FALSE
F3   FALSE FALSE FALSE FALSE FALSE FALSE FALSE FALSE FALSE FALSE FALSE

$lbound: No lower bounds assigned.

$ubound: No upper bounds assigned.
```

The S-matrix with variances and covariances will also be created in steps. It actually consists of a variance covariance matrix of the observed variables and a variance covariance matrix of the factors. First, I am going to create the matrix with the residual variances of the observed variables. These are represented by θ's in Fig. 6.1. The matrix theta is an 8 by 8 matrix, with freely estimated parameters on its diagonal. As there are no residual covariances in the model, all off-diagonal elements are fixed at zero. First, I create an 8 by 8 matrix with zero's, and then I add the vector with the information about the residual variance on its diagonal.

```
theta <- matrix(0,nrow = 8,ncol = 8)
diag(theta) <- c("0.1*t11","0.1*t22","0.1*t33","0.1*t44",
                 "0.1*t55","0.1*t66","0.1*t77","0.1*t88")
```

The phi matrix contains the variances and covariances of the factors. For identification, the factor variances are fixed at 1. The correlations between the factors are specified off-diagonal.

```
phi <- matrix(
         c(1,"0.1*phi21","0.1*phi31",
           "0.1*phi21",1,"0.1*phi32",
           "0.1*phi31","0.1*phi32",1),
         nrow = 3,
         ncol = 3)
```

The function bdiagMat() creates the larger S-matrix from the theta and phi matrices. By using the as.MxMatrix() function on this S-matrix, the matrices with labels, starting values and free/fixed elements to be used by OpenMx are created.

```
S <- bdiagMat(list(theta, phi))
S <- as.mxMatrix(S)

dimnames(S) <- list(
c("face","pict","faci","sens","chen","blen","emma","emre","F1
","F2","F3"),
c("face","pict","faci","sens","chen","blen","emma","emre","F1
","F2","F3"))
```

The resulting S-matrices look like below.

```
FullMatrix 'S'

$labels
       face  pict  faci  sens  chen  blen  emma  emre  F1       F2       F3
face  "t11" NA    NA    NA    NA    NA    NA    NA    NA       NA       NA
pict  NA    "t22" NA    NA    NA    NA    NA    NA    NA       NA       NA
faci  NA    NA    "t33" NA    NA    NA    NA    NA    NA       NA       NA
sens  NA    NA    NA    "t44" NA    NA    NA    NA    NA       NA       NA
chen  NA    NA    NA    NA    "t55" NA    NA    NA    NA       NA       NA
blen  NA    NA    NA    NA    NA    "t66" NA    NA    NA       NA       NA
emma  NA    NA    NA    NA    NA    NA    "t77" NA    NA       NA       NA
emre  NA    NA    NA    NA    NA    NA    NA    "t88" NA       NA       NA
F1    NA    NA    NA    NA    NA    NA    NA    NA    NA       "phi21"  "phi31"
F2    NA    NA    NA    NA    NA    NA    NA    NA    "phi21"  NA       "phi32"
F3    NA    NA    NA    NA    NA    NA    NA    NA    "phi31"  "phi32"  NA

$values
     face pict faci sens chen blen emma emre  F1  F2  F3
face 0.1  0.0  0.0  0.0  0.0  0.0  0.0  0.0 0.0 0.0 0.0
pict 0.0  0.1  0.0  0.0  0.0  0.0  0.0  0.0 0.0 0.0 0.0
faci 0.0  0.0  0.1  0.0  0.0  0.0  0.0  0.0 0.0 0.0 0.0
sens 0.0  0.0  0.0  0.1  0.0  0.0  0.0  0.0 0.0 0.0 0.0
chen 0.0  0.0  0.0  0.0  0.1  0.0  0.0  0.0 0.0 0.0 0.0
blen 0.0  0.0  0.0  0.0  0.0  0.1  0.0  0.0 0.0 0.0 0.0
emma 0.0  0.0  0.0  0.0  0.0  0.0  0.1  0.0 0.0 0.0 0.0
emre 0.0  0.0  0.0  0.0  0.0  0.0  0.0  0.1 0.0 0.0 0.0
F1   0.0  0.0  0.0  0.0  0.0  0.0  0.0  0.0 1.0 0.1 0.1
F2   0.0  0.0  0.0  0.0  0.0  0.0  0.0  0.0 0.1 1.0 0.1
F3   0.0  0.0  0.0  0.0  0.0  0.0  0.0  0.0 0.1 0.1 1.0

$free
      face  pict  faci  sens  chen  blen  emma  emre    F1    F2    F3
face  TRUE  FALSE FALSE FALSE FALSE FALSE FALSE FALSE FALSE FALSE FALSE
pict  FALSE TRUE  FALSE FALSE FALSE FALSE FALSE FALSE FALSE FALSE FALSE
faci  FALSE FALSE TRUE  FALSE FALSE FALSE FALSE FALSE FALSE FALSE FALSE
sens  FALSE FALSE FALSE TRUE  FALSE FALSE FALSE FALSE FALSE FALSE FALSE
chen  FALSE FALSE FALSE FALSE TRUE  FALSE FALSE FALSE FALSE FALSE FALSE
blen  FALSE FALSE FALSE FALSE FALSE TRUE  FALSE FALSE FALSE FALSE FALSE
emma  FALSE FALSE FALSE FALSE FALSE FALSE TRUE  FALSE FALSE FALSE FALSE
emre  FALSE FALSE FALSE FALSE FALSE FALSE FALSE TRUE  FALSE FALSE FALSE
F1    FALSE FALSE FALSE FALSE FALSE FALSE FALSE FALSE FALSE TRUE  TRUE
F2    FALSE FALSE FALSE FALSE FALSE FALSE FALSE FALSE TRUE  FALSE TRUE
F3    FALSE FALSE FALSE FALSE FALSE FALSE FALSE FALSE TRUE  TRUE  FALSE

$lbound: No lower bounds assigned.

$ubound: No upper bounds assigned.
```

Now the required matrices for the Stage 2 analysis are created, the model can be fit to the pooled matrix from Stage 1. As the heterogeneity seems to be substantial, I will fit the model to the Stage 1 matrix from the random effects analysis. The `tssem()` function distils the averaged correlation matrix and the asymptotic variance covariance matrix from the Stage 1 object `stage1random`. As with the path model I used the `diag.constraints = TRUE` and I asked for likelihood based confidence intervals around the parameter estimates.

```
stage2_random <- tssem2(stage1random, Amatrix=A, Smatrix=S,
        Fmatrix=F, diag.constraints=TRUE, intervals="LB")
```

The output can be viewed using the `summary()` function.

```
95% confidence intervals: Likelihood-based statistic
Coefficients:
        Estimate Std.Error  lbound  ubound
L11      0.53025        NA 0.49697 0.56407
L21      0.51982        NA 0.48654 0.55357
L31      0.57671        NA 0.53907 0.61460
L41      0.58797        NA 0.55268 0.62363
L52      0.67185        NA 0.61658 0.72768
L62      0.63164        NA 0.57877 0.68495
L73      0.65046        NA 0.60409 0.69766
L83      0.68395        NA 0.63626 0.73287
t11      0.71883        NA 0.68182 0.75302
t22      0.72979        NA 0.69356 0.76327
t33      0.66741        NA 0.62227 0.70940
t44      0.65429        NA 0.61108 0.69455
t55      0.54862        NA 0.47047 0.61983
t66      0.60103        NA 0.53084 0.66503
t77      0.57691        NA 0.51325 0.63508
t88      0.53222        NA 0.46289 0.59518
phi21    0.60974        NA 0.55321 0.67299
phi31    0.62987        NA 0.57794 0.68595
phi32    0.66528        NA 0.59029 0.74944


Goodness-of-fit indices:
                                                   Value
Sample size                                   10564.0000
Chi-square of target model                       42.2013
DF of target model                               17.0000
p value of target model                           0.0006
Number of constraints imposed on "Smatrix"        8.0000
DF manually adjusted                              0.0000
Chi-square of independence model               2486.1537
DF of independence model                         28.0000
RMSEA                                             0.0118
SRMR                                             0.0257
TLI                                              0.9831
CFI                                              0.9897
AIC                                              8.2013
BIC                                            -115.3073
OpenMx status1: 0 ("0" or "1": The optimization is considered
fine.
Other values indicate problems.)
```

The 8 by 8 pooled correlation matrix on which the model is fitted contains 28 correlation coefficients. The model contains 8 factor loadings, 8 residual variances, and 3 factor covariances (factor variances were fixed at 1), which sums up to 19 parameters. However, because during estimation the 8 diagonal elements of the estimated covariance are constrained to be 1, this reduces the number of parameters by 8. Degrees of freedom are therefore equal to $28 - 19 + 8 = 17$. The model does not fit exactly, as the chi-square is significant ($\chi^2_{(17)} = 42.20$, $p < 0.05$).

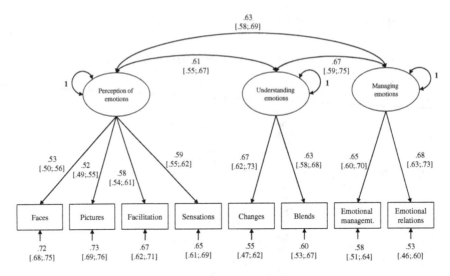

Fig. 6.2 Factor model on the MSCEIT with parameter estimates and 95 % confidence intervals

The RMSEA value of 0.012 indicates close approximate fit, and the CFI of 0.99 also indicated satisfactory fit of the model. The parameter estimates with the confidence intervals could therefore be interpreted. All factor loadings are positive, larger than 0.50, and significantly larger than zero. The correlations between the three factors is substantial (0.62, 0.64 and 0.67), but not so large that some factors may be redundant. Figure 6.2 shows the graphical model with the parameter estimates.

As long as there are no mediating variables in the model, an alternative to using the argument `diag.constraints = TRUE` in the `tssem2()` function is to use `diag.constraints = FALSE` (or to leave out this argument). This will lead to the same fit results and parameter estimates, but the way the analysis is performed is different. Without the diagonal constraints, the diagonals are totally left out of the analysis (the diagonal entries are not counted as observations), and no residual variances (Θ) are estimated. Because a correlation matrix is analyzed, we know that the total variance of each indicator equals 1. The residual variances can therefore be calculated from the matrix with estimated factor loadings (Λ) and matrix with estimated factor variances and covariances (Φ) using $\Theta = I - \text{diag}(\Lambda \Phi \Lambda^T)$, where I is an 8 by 8 identity matrix.

Acknowledgment I am grateful to Dr. Huiyong Fan and the co-authors of his study for sharing the collected correlation matrices with me. Indirectly, I am also grateful to the authors of the primary studies, who were willing to share their correlation matrices with them. I would like to stress that the ownership of these data belongs to the primary authors. I put the file with the correlation matrices on my Web site for educational purposes. Any part of these data should not be used for other purpose than research or education, without prior written permission from these primary authors or Dr. Huiyong Fan.

References

Fan, H., Jackson, T., Yang, X., Tang, W., & Zhang, J. (2010). The factor structure of the Mayer-Salovey-Caruso Emotional Intelligence Test V2.0 (MSCEIT): A meta-analytic structural equation modeling approach. *Personality and Individual Differences, 48*(7), 781–785.

Salovey, P., & Mayer, J. D. (1989). Emotional intelligence. *Imagination, Cognition and Personality, 9*(3), 185–211.

Appendix A
Model Implied Covariance Matrix of the Example Path Model

$$\begin{bmatrix} \psi_{11} & & & \\ \psi_{21} & \psi_{22} & & \\ \beta_{31}\psi_{11} + \beta_{32}\psi_{21} & \beta_{31}\psi_{21} + \beta_{32}\psi_{22} & & \\ \beta_{31}\beta_{43}\psi_{11} + \beta_{32}\beta_{43}\psi_{21} & \beta_{31}\beta_{43}\psi_{21} + \beta_{32}\beta_{43}\psi_{22} & (\beta_{31}\psi_{11} + \beta_{32}\psi_{21})\beta_{31} + (\beta_{31}\psi_{21} + \beta_{32}\psi_{22})\beta_{32} + \psi_{33} & \end{bmatrix}$$

And the model implied variance of Variable 4 (didn't fit in the matrix above)

$$\left[(\beta_{31}\beta_{43}\psi_{11} + \beta_{32}\beta_{43}\psi_{21})\beta_{43}\beta_{31} + (\beta_{31}\beta_{43}\psi_{21} + \beta_{32}\beta_{43}\psi_{22})\beta_{43}\beta_{32} + \beta_{43}^2\psi_{33} + \psi_{44} \right]$$

© The Author(s) 2015
S. Jak, *Meta-Analytic Structural Equation Modelling*,
SpringerBriefs in Research Synthesis and Meta-Analysis,
DOI 10.1007/978-3-319-27174-3

Appendix B
Fitting a Path Model to a Covariance Matrix with OpenMx

In this appendix I explain how to fit the path model from Chap. 1 to a covariance matrix. I assume that you understand the basics of R. There exist different approaches to fit models using OpenMx. I will use the matrix based approach. Another way would be to use the "path model specification".

To get started with OpenMx you first download the package 'OpenMx'. With the command:

```
source('http://openmx.psyc.virginia.edu/getOpenMx.R')
```

the latest version of the package 'OpenMx' will be downloaded from the OpenMx website. You have to do this only the first time you use OpenMx (on a specific computer), to add the package to the R library. To activate the package into the current R workspace, you type

```
require(OpenMx)
```

A script to fit a path model with OpenMx consists of four steps. First, the observed covariance matrix has to be specified in R. Second, the model has to be specified. Third, we fit the model to the observed covariance matrix, by submitting both model and covariance matrix to OpenMx. Finally, the output needs to be retrieved from the object where all results are stored.

The script below fits the path model from Chap. 1 to a covariance matrix. All commands will be explained afterwards.

© The Author(s) 2015
S. Jak, *Meta-Analytic Structural Equation Modelling*,
SpringerBriefs in Research Synthesis and Meta-Analysis,
DOI 10.1007/978-3-319-27174-3

```
source('http://openmx.psyc.virginia.edu/getOpenMx.R')

require(OpenMx)

# observed covariance matrix

obsnames <- c("pos","neg","enga","achiev")
obslabels <- list(obsnames,obsnames)


obscov <- matrix(c(.81,-.36,.63,.14,
                   -.36,1.21,-.60,-.33,
                   .63,-.60,1.69,.50,
                   .14,-.33,.50,1.44),
               nrow = 4, ncol = 4,
               dimnames = obslabels)
# define the model

title <- "Path model"

obs <-  mxData(observed = obscov, type = "cov", numObs =
104)

matrixB <-    mxMatrix(
              type = "Full",
              nrow = 4,
              ncol = 4,
              free = c(FALSE,FALSE,FALSE,FALSE,
                       FALSE,FALSE,FALSE,FALSE,
                       TRUE,TRUE,FALSE,FALSE,
                       FALSE,FALSE,TRUE,FALSE),

              values = c(0,0,0,0,
                         0,0,0,0,
                         1,1,0,0,
                         0,0,1,0),

              labels = c(NA,NA,NA,NA,
                         NA,NA,NA,NA,
                         "b31","b32",NA,NA,
                         NA,NA,"b43",NA),

              byrow = TRUE,
              name = "B",
              dimnames = obslabels)

matrixP <-    mxMatrix(
              type = "Symm",
              nrow = 4,
              ncol = 4,
```

```
                    free = c(TRUE,
                             TRUE,TRUE,
                             FALSE,FALSE,TRUE,
                             FALSE,FALSE,FALSE,TRUE),

                    values = c(1,
                               .5,1,
                               0,0,1,
                               0,0,0,1),

                    byrow = TRUE,
                    name = "P",
                    dimnames = obslabels)
matrixI <-      mxMatrix(
                    type = "Iden",
                    nrow = 4,
                    ncol = 4,
                    name = "I")

Ind_pos <- mxAlgebra(expression = "b31"*"b43",
                          name = "Ind_pos")
Ind_neg <- mxAlgebra(expression = "b32"*"b43",
                          name = "Ind_neg")

conf <- mxCI(c("B","P","Ind_pos","Ind_neg"),interval = .95)

algebraS <- mxAlgebra(expression =
                          solve(I-B) %*% P %*% t(solve(I-B)),
                          name = "Sigma", dimnames = obslabels)

exp <- mxExpectationNormal(covariance="Sigma")

fit <- mxFitFunctionML()

pathmodel<- mxModel(title,obs,matrixB,matrixP,matrixI,
                        Ind_pos,Ind_neg,conf,algebraS,exp,fit)

pathmodelOut <- mxRun(pathmodel, intervals = TRUE)

# retrieve the output

summary(pathmodelOut)

pathmodelOut$B@values
pathmodelOut$P@values
```

We start by defining the observed covariance matrix and the labels (names) of the associated observed variables. It is required to provide these labels with the input matrix. The labels are given as a list with two elements, one vector of row names and one vector of column names. First we created the object `obsnames` with the names of the variables.

```
obsnames <- c("pos","neg","enga","achiev")
obslabels <- list(obsnames,obsnames)
```

The observed covariance matrix is stored in the object `obscov`, by creating a matrix with the values of the elements, number of rows, number of columns, and the name vectors of the two dimensions.

```
obscov <- matrix()
```

To check whether you successfully specified the observed covariance matrix, check the results by typing `obscov` in the R console. And check, for example, whether the matrix is indeed symmetrical by typing `obscov == t(obscov)`.

The next step is to specify the model that has to be fitted to the observed data. The OpenMx package has several functions that we will use. The main functions are `mxModel()` and `mxRun()`. `mxModel()` is a 'container-function' that results in an object that contains all the information needed to fit the model. The model is actually fitted with the `mxRun()` function. All parts of the model fitting process are first created as separate objects, and then stored in another object using the `mxModel()` function. The different objects are:

– A title
– The data (e.g. observed covariance matrix)
– The matrices containing the model parameters
– The expected (model implied) covariance matrix and fit-function

The code

```
title <- "Path model"
```

defines an object with the title, do not forget the " " to make the object of the type character (i.e., so R knows that title is a line of text, not a number).

```
obs <-  mxData(observed = obscov, type = "cov",
               numObs = 104)
```

This line creates the object `obs` to store the outcome of the function `mxData()`. The function `mxData()` has three arguments: (1) `observed =` for the observed matrix (that you specified previously), (2) `type =` for the type of the matrix (`"cor"` for correlation matrix, and `"cov"` for covariance matrix), and (3) `numObs =` for the number of observations, the sample size.

The model implied covariance matrix of a path model is as follows:

$$\Sigma = (I - B)^{-1} * \Psi * (I - B)^{-1t}, \qquad (B.1)$$

where Σ is the matrix with the resulting model implied variances and covariances, I is an identity matrix, B is a matrix containing the direct effects, and Ψ is a matrix containing variances and covariances. In openMx we will denote Σ, B, Ψ and I with respectively Sigma, B, P and I.

```
matrixB <-    mxMatrix(
              type = "Full",
              nrow = 4,
              ncol = 4,

free = c(FALSE, FALSE, FALSE, FALSE,
         FALSE, FALSE, FALSE, FALSE,
         TRUE, TRUE, FALSE, FALSE,
         FALSE, FALSE, TRUE, FALSE),

values = c(0,0,0,0,
           0,0,0,0,
           1,1,0,0,
           0,0,1,0),


labels = c(NA, NA, NA, NA,
           NA, NA, NA, NA,
           "b31", "b32", NA, NA,
           NA, NA, "b43", NA),

byrow = TRUE,
name = "B",
dimnames = obslabels)
```

Matrix B contains the parameters for the direct effects. Matrix B is constructed using the function `mxMatrix()` with several arguments. B is a full matrix (`type = "Full"`), with numbers of rows and columns equal to the number of observed variables. The argument

```
free = c(FALSE, FALSE, FALSE, FALSE,
         FALSE, FALSE, FALSE, FALSE,
         TRUE, TRUE, FALSE, FALSE,
         FALSE, FALSE, TRUE, FALSE),
```

is a vector (in matrix shape) specifying which elements in matrix B should be estimated (`TRUE`) and which should not be estimated (`FALSE`). Both the rows and the columns of the B matrix are associated with the (four) observed variables.

You need to be careful to specify these direct effects correctly. It may help to realize that the columns are associated with the independent variables, and the rows with the dependent variables. Another way to think about it is to formulate in terms of regression: when Variable 4 is regressed on Variable 1, you specify TRUE in element B 4 1. The diagonal of B is always FALSE, as a variable cannot be regressed on itself. The argument

```
values = c(0,0,0,0,
           0,0,0,0,
           1,1,0,0,
           0,0,1,0),
```

is a vector of values. These values are fixed values for the fixed (FALSE) elements (usually zero), or start values for the parameters to be estimated (TRUE). To specify a start value for the direct effect of Variable 1 on Variable 4, or regression of Variable 4 on Variable 1 (β_{41}), we put a 1 (or some other value) in row 4, column 1 of the B matrix. The parameters can be given labels, which are for example needed if you want to calculate indirect effects or add equality constraints later on, by adding:

```
labels = c(NA,NA,NA,NA,
           NA,NA,NA,NA,
           "b31","b32",NA,NA,
           NA,NA,"b43",NA),
```

The argument:

```
byrow = TRUE,
name = "B",
dimnames = obslabels
```

should not be forgotten, because the matrix should be filled with elements by row and not by column (which is the default). The matrix is given a name ("B") that can be used within other parts of the mxModel. The last argument provides labels to the rows and columns of the matrix.

```
matrixP <-    mxMatrix(
              type = "Symm",
              nrow = 4,
              ncol = 4,
              free = c(TRUE,
                       TRUE,TRUE,
                       FALSE,FALSE,TRUE,
                       FALSE,FALSE,FALSE,TRUE),

              values = c(1,
                         .5,1,
                         0,0,1,
                         0,0,0,1),

              byrow = TRUE,
              name = "P",
              dimnames = obslabels)
```

Matrix P is also an mxMatrix and contains the variances and covariances between variables (or between disturbances for endogenous variables). P is a symmetrical matrix (type = "Symm") with the same dimensions as the B matrix (i.e., the number of observed variables). The free elements in the P matrix are provided in a symmetrical matrix with FALSE for fixed elements, and TRUE for free to be estimated elements. In this example there are TRUE's on the diagonal, meaning that the variances of the exogenous variables and the disturbance variances of the endogenous variables are free to be estimated. Off diagonal there is only one TRUE, for the covariance between the two exogenous variables (The double headed arrow between "Positive relations" and "Negative relations" in Chap. 1).

```
matrixI <-    mxMatrix(
              type = "Iden",
              nrow = 4,
              ncol = 4,
              name = "I")
```

Matrix I is an identity matrix (type = "Iden"), with the same dimensions as the B and P matrices (i.e., the number of observed variables). It needs fewer arguments, because all elements of an identity matrix are fixed.

```
Ind_pos <- mxAlgebra(expression = "b31"*"b43",
                     name = "Ind_pos")
Ind_neg <- mxAlgebra(expression = "b32"*"b43",
                     name = "Ind_neg")
```

The indirect effects from Negative and Positive interactions on Achievement, through Engagement, are calculated in an mxAlgebra() function, by referring to the labels of the two direct effects that make up the indirect effect.

```
conf <- mxCI(c("B","P","Ind_pos","Ind_neg"),
             interval = .95)
```

With the mxCI() function we ask for 95 % likelihood based confidence intervals for all elements in the B and P matrices, and the indirect effects.

```
algebraS <- mxAlgebra(expression =
            solve(I-B) %*% P %*%  t(solve(I-B)),
            name = "Sigma", dimnames = obslabels)
```

The model implied covariance matrix is defined in algebraS, with the mxAlgebra() function, using the matrices that have been defined before in the expression of the path model. We name this model implied matrix "Sigma". The model implied covariance matrix is given the same labels as the observed covariance matrix through dimnames = obslabels.

```
exp <- mxExpectationNormal(covariance="Sigma")

fit <- mxFitFunctionML()
```

The `mxMLObjective()` function has as arguments the model implied covariance matrix. The `mxFitFunctionML()` function does not need arguments, but has to be added to the mxModel to indicate that we want to use the maximum likelihood fit function. Now, all separate elements of an mxModel are created, and we can build the actual mxModel, calling it 'pathmodel':

```
pathmodel <- mxModel(title,obs,matrixB,matrixP,
            matrixI,Ind_pos,Ind_neg,conf,algebraS,exp,fit)
```

We actually fit ("run") the model by specifying `mxRun(pathmodel)` and store the output in 'pathmodelOut':

```
pathmodelOut <- mxRun(pathmodel, intervals = TRUE)
```

The `intervals = TRUE` argument can be used to specify whether the confidence intervals should be estimated or not. For very large models it may take a long time to estimate the intervals so the argument may be set to FALSE in some cases.

In order to get information about the model fit and parameter estimates, we can ask for a summary of the output:

```
summary(pathmodelOut)
```

In the summary, one will see the observed covariance matrix, the parameter estimates with standard errors and some of the fit results. This model has 2 degrees of freedom, and a chi-square of 2.538.

Information about the parameter estimates only, in matrix shape can be obtained with:

```
pathmodelOut$B$values
pathmodelOut$P$values
```

Appendix C
Model Implied Covariance Matrix of the Example Factor Model

$$
\begin{bmatrix}
\phi_{11} + \theta_{11} & & & & \\
\lambda_{21}\phi_{11} & \lambda_{21}^2\phi_{11} + \theta_{22} & & & \\
\lambda_{31}\phi_{11} & \lambda_{21}\phi_{11}\lambda_{31} & \lambda_{31}^2\phi_{11} + \theta_{33} & & \\
\phi_{21} & \lambda_{21}\phi_{21} & \lambda_{31}\phi_{21} & \phi_{22} + \theta_{44} & \\
\lambda_{52}\phi_{21} & \lambda_{21}\phi_{21}\lambda_{52} & \lambda_{31}\phi_{21}\lambda_{52} & \lambda_{52}\phi_{22} & \lambda_{52}^2\phi_{22} + \theta_{55}
\end{bmatrix}
$$

© The Author(s) 2015
S. Jak, *Meta-Analytic Structural Equation Modelling*,
SpringerBriefs in Research Synthesis and Meta-Analysis,
DOI 10.1007/978-3-319-27174-3

Appendix D
Fitting a Factor Model to a Covariance Matrix with OpenMx

The openMx script for fitting a factor model resembles the script for fitting a path model. The biggest difference is that the model of Σ is now a factor model:

$$\Sigma = \Lambda\Phi\Lambda' + \Theta, \tag{D.1}$$

where Λ is a full matrix with factor loadings, Φ is a symmetric matrix with variances and covariances of the common factors, and Θ is a diagonal matrix with variances (or sometimes a symmetric matrix with covariances) of the residual factors.

The script below fits the two-factor model depicted in Chap. 1, to the observed covariance matrix.

```
# observed covariance matrix

obsnames <- c("with","somat","anxiety","delinq","aggres")
factornames <- c('Internalizing','Externalizing')

obslabels <- list(obsnames,obsnames)
factorlabels <- list(factornames,factornames)
lambdalabels <- list(obsnames,factornames)

CBCLcov <- matrix(c(12.554,  6.306,11.147,  2.846,12.437,
                     6.306,10.057,  9.642,  2.090,  9.679,
                    11.147,  9.642,26.018,  4.836,22.199,
                     2.846,  2.090,  4.836,  3.718,  9.962,
                    12.437,  9.679,22.199,9.962,  51.020),
             nrow = 5, ncol = 5,
             dimnames = obslabels)
```

© The Author(s) 2015
S. Jak, *Meta-Analytic Structural Equation Modelling*,
SpringerBriefs in Research Synthesis and Meta-Analysis,
DOI 10.1007/978-3-319-27174-3

```
# Model

title <- "Factor model CBCL"

obs <-  mxData(observed = CBCLcov, type = "cov",
               numObs = 155)

matrixL <-  mxMatrix(
               type = "Full",
               nrow = 5,
               ncol = 2,
               free = c(FALSE,FALSE,
                        TRUE,FALSE,
                        TRUE,FALSE,
                        FALSE,FALSE,
                        FALSE,TRUE),

               values = c(1,0,
                          1,0,
                          1,0,
                          0,1,
                          0,1),

               byrow = TRUE,
               name = "L",
               dimnames = list(obsnames,factornames))


matrixF <-  mxMatrix(
               type = "Symm",
               nrow = 2,
               ncol = 2,
               free = c(TRUE,
                        TRUE,TRUE),

               values = c(1,
                          .5,1),
               byrow = TRUE,
               name = "F",
               dimnames = factorlabels)
```

```
matrixT <-  mxMatrix(
               type = "Diag",
               nrow = 5,
               ncol = 5,

               free = c(TRUE,TRUE,TRUE,TRUE,TRUE),

               values = c(1,1,1,1,1),

               byrow = TRUE,
               name = "T",
               dimnames = obslabels)

conf <- mxCI(c("L","F","T"),interval = .95)

algebraS <- mxAlgebra(expression = L%*%F%*%t(L) + T, name =
                      "Sigma", dimnames = obslabels)

exp <- mxExpectationNormal(covariance="Sigma")

fit <- mxFitFunctionML()

CBCLmodel <- mxModel(title,obs,matrixL,matrixF,matrixT,
                     conf,algebraS,exp,fit)
CBCLmodelOut <- mxRun(CBCLmodel)

# retrieve the output

summary(CBCLmodelOut)

CBCLmodelOut$L$values
CBCLmodelOut$F$values
CBCLmodelOut$T$values
```

The first part of the script, where the observed covariance matrix is created, is not different from when fitting a path model. Differences are present in the matrices that are used. The matrices involved in a factor model are Λ, Φ and Θ. In openMx we will use L, F, and Q to denote Λ, Φ, and Θ, respectively.

To facilitates reading the results in the L, F and Q matrix, we now need both the names of the observed variables and the common factors. Therefore, we also created an object with the names of the common factors:

```
factornames <- c('Internalizing','Externalizing')
```

And we create the lists with the labels for the matrices in the factormodel:

```
obslabels <- list(obsnames,obsnames)
factorlabels <- list(factornames,factornames)
lambdalabels <- list(obsnames,factornames)
```

The labels of the Lambda matrix involve both the names of the observed variables (the rows) and the names of the common factors (the columns).

Matrix L contains the factor loadings. Factor loadings are the regression coefficients for the regressions of the indicator variables on the common factors (i.e., the effects of the factors on the indicator variables). L is always a full matrix, with the number of rows equal to the number of indicators and the number of columns equal to the number of common factors.

The argument

```
free = c(FALSE,FALSE,
         TRUE,FALSE,
         TRUE,FALSE,
         FALSE,FALSE,
         FALSE,TRUE),
```

specifies which factor loadings should be estimated and which are fixed. To correctly fix and free elements, it may help to think of the columns as being associated with the common factors and the rows as being associated with the indicators. For example if indicator number three loads on the first factor (or, the third indicator variable regresses on the first factor), we specify TRUE for element (3,1). Start values and fixed values are provided in the same way as with a path model. In this example, the first factor loading per factor is fixed to 1 to give the factors a scale. The alternative would be to fix the variances of the factors to 1.

```
values = c(1,0,
           1,0,
           1,0,
           0,1,
           0,1),
```

For start values of the free loadings, we used 1.

Matrix F contains the variances and covariances of the common factors. As it is a covariance matrix, it is always a symmetric matrix. Its dimensions are equal to the number of common factors. Because we fixed one factor loading per factor at 1 in this example, the factor variances can be estimated. The TRUE at the off diagonal indicates that the covariance between the common factors is free to be estimated. So there is a TRUE for all elements in F. Start values for the elements in F are also given.

```
matrixF <-    mxMatrix(
              type = "Symm",
              nrow = 2,
              ncol = 2,
              free = c(TRUE,
                          TRUE,TRUE),

              values = c(1,
                          .5,1),

              byrow = TRUE,
              name = "F",
              dimnames = factorlabels)
```

Matrix T contains the variances of the residual factors. As there are no covariances between the residual factors, matrix T is a diagonal matrix, with dimensions equal to the number of indicators. In this example, all residual variance should be estimated, so we provided all TRUE's in the "free" argument.

```
matrixT <-    mxMatrix(
              type = "Diag",
              nrow = 5,
              ncol = 5,
              free = c(TRUE,TRUE,TRUE,TRUE,TRUE),
              values = c(1,1,1,1,1),
              byrow = TRUE,
              name = "T",
              dimnames = obslabels)
```

We ask for likelihood based confidence intervals for all parameter estimates:

```
conf <- mxCI(c("L","F","T"),interval = .95)
```

The expression for the expected covariance matrix is now a factor model:

```
algebraS <- mxAlgebra(expression = L%*%F%*%t(L) + Q,
              name = "Sigma", dimnames = obslabels)

exp <- mxExpectationNormal(covariance="Sigma")

fit <- mxFitFunctionML()
```

Finally, all the elements of the model are collected in the mxModel function, and run with the mxRun function.

```
CBCLmodel <- mxModel(title,obs,matrixL,matrixF,
                 matrixT,conf,algebraS,exp,fit)

CBCLmodelOut <- mxRun(CBCLmodel)
```

The results of the analyses, and the estimates of the L, F and T matrices can be obtained with:

```
summary(CBCLmodelOut)

CBCLmodelOut$L$values
CBCLmodelOut$F$values
CBCLmodelOut$T$values
```

Printed in the United States
By Bookmasters